ON
TRAILS

ROBERT
MOOR

SIMON & SCHUSTER

New York   London   Toronto   Sydney   New Delhi

The path is made in the walking of it.

—ZHUANGZI

CONTENTS

ON TRAILS

PROLOGUE

ONCE, YEARS AGO, I left home looking for a grand adventure and spent five months staring at mud. It was the spring of 2009, and I had set out to walk the full length of the Appalachian Trail from Georgia to Maine. My departure date was timed so that I would transition seamlessly from a mild southern spring to a balmy northern summer, but for some reason the warmth never arrived. It stayed cool that year, rained often. Newspapers likened it to the freak summer of 1816, when cornfields froze to their roots, pink snow fell over Italy, and a young Mary Shelley, locked up in a gloomy villa in Switzerland, began to dream of monsters. My memories of the hike consist chiefly of wet stone and black earth. The vistas from many of the mountaintops were blotted out. Shrouded in mist, rain hood up, eyes downcast, mile after mile, month after month, I had little else to do but study the trail beneath my nose with Talmudic intensity.

In his novel *The Dharma Bums*, Jack Kerouac refers to this kind of walking as "the meditation of the trail." Japhy Ryder, a character modeled after the Zen poet Gary Snyder, advises his friend to "walk along looking at the trail at your feet and don't look about and just fall into a trance as the ground zips by." Trails are seldom looked at this intently. When hikers want to complain about a particularly rough stretch of trail, we gripe that we spent the whole day looking down at our feet. We prefer to look up, away, off into the distance. Ideally, a trail should function like a discreet aide, gracefully ushering us through the world while still preserving our sense of agency and independence. Perhaps this is why, for virtually all of literary history, trails have remained in the periphery of our gaze, down at the bottommost edge of the frame: they have been, quite literally, beneath our concern.

As hundreds—and then thousands—of miles of trail passed beneath my eyes, I began to ponder the meaning of this endless scrawl. Who created it? Why does it exist? Why, moreover, does any trail?

Even after I reached the end of the AT, these questions followed me around. Spurred on by them, and sensing in some vague way that they might lead to new intellectual ground, I began to search for the deeper meaning of trails. I spent years looking for answers, which led me to yet bigger questions: Why did animal life begin to move in the first place? How does any creature start to make sense of the world? Why do some individuals lead and others follow? How did we humans come to mold our planet into its current shape? Piece by piece, I began to cobble together a panoramic view of how pathways act as an essential guiding force on this planet: on every scale of life, from microscopic cells to herds of elephants, creatures can be found relying on trails to reduce an overwhelming array of options to a single expeditious route. Without trails, we would be lost.

My quest to find the nature of trails often proved trickier than I had expected. Modern hiking trails loudly announce their presence with brightly painted signs and blazes, but older trails are more inconspicuous. The footpaths of some ancient indigenous societies, like the Cherokee, were no more than a few inches wide. When Europeans invaded North America, they slowly widened parts of the native trail network, first to accommodate horses, then wagons, then automobiles. Now, much of that network is buried beneath modern roadways, though remnants of the old trail system can still be found when you know where—and how—to look.

Other trails are yet more obscure. The trails of some woodland mammals dimple the underbrush so faintly that only an experienced tracker can point them out. Ants nose along chemical pathways that are wholly invisible. (One trick to seeing them, I learned, is to sprinkle the area with lycopodium, the same powder police use to dust for fingerprints.) A few trails are tucked away underground: termites and naked mole-rats carve tunnels through the earth, marking them with traces of pheromones to keep their bearings. Finer still are the tangled neural pathways within a single human brain, which are so multitudinous that even the world's most advanced computers cannot yet map them all. Technology, meanwhile, is busy knitting itself into an intricate network of pathways, dug deep underfoot and strung ethereally overhead, so that information can race across continents.

I learned that the soul of a trail—its *trail-ness*—is not bound up in dirt and rocks; it is immaterial, evanescent, as fluid as air. The essence lies in its function: how it continuously evolves to serve the needs of its users. We tend to glorify trailblazers—those hardy souls who strike out across uncharted territory, both figurative and physical—but followers play an equally important role in creating a trail. They shave off unnecessary bends and brush away obstructions, improving the trail with each trip. It is thanks to the actions of these walkers that the trail becomes, in the words of Wendell Berry, "the perfect adaptation, through experience and familiarity, of movement to place." In bewildering times—when all the old ways seem to be dissolving into mire—it serves us well to turn our eyes earthward and study the oft-overlooked wisdom beneath our feet.

+

I was ten years old when I first glimpsed that a trail could be something more than a strip of bare dirt. That summer, my parents shipped me off to a small, antiquated summer camp in Maine called Pine Island, where there was no electricity or running water, only kerosene lanterns and cold lake. During the second of my six weeks there, a handful of us boys were loaded into a van and driven many hours away to the base of Mount Washington, for what was to be my first backpacking trip. As a child of the concretized prairies of suburban Illinois, I was apprehensive. The act of lugging a heavy pack through the mountains looked suspiciously like one of those penitent rituals that adults sometimes forced themselves to perform, like visiting distant relatives or eating crusts of bread.

I was wrong, though; it was worse. Our counselors had allotted us three days to climb the eight miles to the top of Mount Washington and back down, which should have been ample time. But the trail was steep, and I was scrawny. My backpack—a heavy, ill-fitting, aluminum-framed Kelty—resembled a piece of full-body orthodontia. After only an hour of climbing the wide rocky trail leading up Tuckerman Ravine, my stiff new leather boots had already begun to blister my toes and rasp the skin from my heels. A hot liquid ache perfused the muscles of my back. When my counselors weren't looking, I made pleading, pained faces at passing strangers, as if this were all part of some elaborate kidnapping. That night, as I lay in my sleeping bag in the lean-to, I considered the logistics of an escape.

On the second morning, a gray rain blew in. Instead of summiting the peak, which our counselors deemed unsafe, we took a long hike around the southern flank of the mountain. We left our packs back at the shelter, each of us carrying only a single water bottle and a pocketful of snacks. Free from the dreaded weight of my pack, warm inside my rubberized rain poncho, I began to enjoy myself. I inhaled the fir-sweet air, exhaled fog. The forest gave off a faint chlorophyllic glow.

We walked in single file, floating through the trees like little ghosts. After an hour or two, we rose above the tree line and entered a realm of lichen-crusted rock and white mist. The trails around the mountain branched and twined. At the juncture with the Crawford Path, one of our counselors announced that we were turning onto a leg of the Appalachian Trail. His tone suggested we were meant to be impressed. I had heard that name before, but I wasn't sure what it meant. The path beneath our feet, he explained, followed the spine of the Appalachians north to Maine and south all the way to the state of Georgia, almost two thousand miles away.

I still recall the tingle of wonder I felt upon hearing these words. The plain-looking trail beneath my feet had suddenly grown to colossal scale. It was as if I had dived down into the camp lake and discovered the slow, undulant vastness of a blue whale. Small as I felt back then, it was a thrill to grasp something so immense, if only by the very tip of its tail.

+

I kept hiking. It got easier—or rather, I got tougher. My pack and boots softened until they slid into place with the dry fluidity of an old baseball glove. I learned to move nimbly beneath a heavy load and push on for hours without breaks. I also came to savor the satisfaction of dropping my pack at the end of a long day: the warm animal weight would fall coolly away, and I would rise from my burden with a weird heliated feeling, as if my toes were merely grazing the dirt.

Hiking proved to be the perfect pastime for a free-floating kid like me. My mother once gave me a leather-bound journal that was meant to have my name embossed in gold along the spine, but instead the printer erroneously engraved the words ROBERT MOON. The mistake was oddly fitting. Growing up, I often felt extraterrestrial. It wasn't that I was lonely or ostracized; I just never felt fully *at home*.

Before I went off to college, no one knew I was gay, and I knew no other gay people. I did my best to blend in. Each year I would dutifully put on a suit and tie for the spring formal, the cotillion, or the prom. I donned athletic uniforms, first-date uniforms, drinking-pilfered-cans-of-Old-Style-in-a-friend's-basement uniforms. All the while, though, part of me wondered: What's the point of this elaborately costumed performance we put on?

In my family I was the youngest child by nearly a decade. My parents, who were already in their forties by the time I was born, granted me an unusual amount of freedom. I could have run wild. Instead, I spent much of my time in my room reading books, which, I discovered, was like running away from home, minus the risk and parental heartache. And so, from the third grade on, I burned through books the way a chain-smoker smokes, picking up one even as I was extinguishing the last.

The book that kicked off my habit in earnest was a flimsy pa-perback copy of *Little House in the Big Woods*. I learned that my home, in northern Illinois, was just a few hundred miles southeast of where the book's author, Laura Ingalls Wilder, was born in 1867. However, her descriptions of the Big Woods of Wisconsin were wholly foreign to me. "As far as a man could go to the north in a day, or a week, or a whole month, there was nothing but woods," she wrote. "There were no houses. There were no roads. There were no people. There were only trees and the wild animals who had their homes among them." I was intoxicated by Ingalls's sense of isolation and self-reliance.

I don't remember how many of the Little House books I read in a row, but it was enough to require an intervention from my teacher, who gently suggested I move on to something else. In the coming years I progressed from *Little House* to *Hatchet* to *Walden* to *A Sand County Almanac* to *Pilgrim at Tinker Creek*. I enjoyed lingering over the minutiae of a life spent outdoors. During my first summer at

Pine Island, I discovered a parallel genre of wilderness adventure books: first the boyish yarns of Mark Twain and Jack London, then the alpine reveries of John Muir, the Antarctic agonies of Ernest Shackleton, and the existential odysseys of Robyn Davidson and Bruce Chatwin.

These two lineages of outdoor writers were roughly divided between those who were deeply rooted to a piece of land and those who were proudly untethered. I preferred the drifters. I held no profound connection to my land, my ancestors, my culture, my community, my gender, or my race. I was raised without religion, and without hatred of religion. My family was diffuse: my parents, two Texans living in the frigid North, were already divorced by the time I was in the first grade; not long after, my two older sisters went away to college and never moved back. A vague restlessness seemed to run in our blood.

Nine months out of every year I drifted through the halls of one academic institution after another, changing costumes, learning new dialects, faking fluency. It was only during the summers, on a series of ever-lengthening sojourns in the wilderness, that I felt wholly natural. I worked my way up from the Appalachians to the mighty Rockies, then to the Beartooths, the Winds, the snowy behemoths of the Alaska Range, and, later, high-altitude peaks ranging from Mexico to Argentina. Up there, far from etiquette or ritual, I could walk unscrutinized, unbound.

For two summers in college I took a job back at Pine Island leading kids on short hikes through the Appalachians. On trips along the AT I would occasionally bump into hikers who were attempting to walk the trail's full length in a single, mammoth, months-long effort. These "thru-hikers" were easy to spot: They introduced themselves with odd "trail names," ate ravenously, and walked with a light, lupine gait. I was intimidated by them, but also envious. They resembled the rock musicians of an idealized past—the same long hair, the same wild beards, the same wasted physiques, the same esoteric argot, the same peripatetic lifestyle, the same faint, vain awareness of being, in a way, heroic.

I sometimes talked with these thru-hikers, plying them with chunks of cheese or handfuls of candy. I remember one old man who had hiked the whole trail in a Scottish kilt and sandals, and a young man who carried no tent, but a full feather pillow. A few of them proselytized zealously for one church or another, while others spoke of preparing for a looming ecological apocalypse. Many of the people I talked to were between jobs, between schools, or between marriages. I met soldiers returning from war and people recovering from a death in the family. Certain stock phrases were repeated. "I needed some time to clear my head," they said, or "I knew this might be my last chance." One summer during college, I told a young thru-hiker that I hoped to make an attempt someday. "Drop out," he told me flatly. "Do it now."

+

I did not drop out. I was too careful for that. In 2008 I moved to New York, where I worked a series of low-paying jobs. In my free time I planned my thru-hike. I read guidebooks and online message boards, drew up tentative itineraries. Less than a year later I was ready to embark.

Unlike many people, I had no clear impetus for going on a long hike, no inciting incident. I wasn't grieving a death or recovering from drug addiction. I wasn't fleeing anything. I had never been to war. I wasn't depressed. I was maybe only a little insane. My thru-hike was not an attempt to find myself, find peace, or find God.

Perhaps, as they say, I simply needed some time to clear my head; perhaps I knew this might be my last chance. Both were mostly true, as clichés often are. I also wanted to find out what it would be like to spend months on end in the wilderness, to live in a prolonged state of freedom. But more than that, I think I wanted to answer a challenge that had loomed over me since childhood. When I was small and frail, hiking the whole trail had seemed a herculean task. As I grew, its impossibility became precisely its appeal.

+

Over the years, I had picked up some useful tips from the thru-hikers I'd met. Above all, I knew that weight was the enemy of a successful thru-hike, so I retired my trusty old pack and invested in a new ultralight one. Then I traded in my bulky tent for a hammock, bought an airy goose-down sleeping bag, and exchanged my leather boots for a pair of trail running shoes. I pared my medkit down to a few anti-diarrheal pills, some iodine swabs, a thumb-sized roll of duct tape, and a safety pin. I replaced my white gas stove with one made out of two aluminum Coke cans, which weighed practically nothing. When I crammed all of my gear into my new pack and lifted it for the first time, I was amazed and slightly terrified. It seemed too insubstantial to house, clothe, and feed a human for five months.

So I wouldn't be forced to live off an anemic diet of instant ramen and freeze-dried mashed potatoes, I began cooking heaping piles of nutritious slop (beans and brown rice, quinoa, couscous, whole wheat pasta with tomato sauce) and dehydrating them. I poured sparing amounts of olive oil and hot sauce into small plastic bottles. I filled plastic baggies with baking soda, Gold Bond, vitamins, and painkillers. I divided all of the supplies up into roughly five-day increments and packed them into fourteen cardboard boxes. Into each box, I also placed a chapbook of poetry or a heftier paperback novel that I had cut into slimmer volumes using a straight razor and packing tape.

I addressed these boxes to post offices along the trail—towns with names like Erwin, Hiawassee, Damascus, Caratunk, and (my

favorite) Bland—and left them with my roommate to mail on specified dates. I quit my job. I sublet my apartment. I sold or gave away
everything I could spare. Then, on a cold day in March, I flew down
to Georgia.

+

On the summit of Springer Mountain, the trail's southern terminus, I
was greeted by an old man who called himself Many Sleeps, a moniker he had reportedly earned while completing one of the slowest
thru-hikes ever recorded. With his droopy eyes and long white beard,
he looked like a nylon-clad Rip Van Winkle.

In his hand he held a clipboard. His job was to collect information
from all the passing thru-hikers. He told me it had been a busy year:
twelve thru-hikers had registered with him that day, and thirty-seven
the day before. In total that spring, almost fifteen hundred people
would set out from Springer aiming for Maine, though scarcely a
quarter of them would make it.

There on the mountaintop, before starting my long-awaited hike,
I paused to admire the land below: swells of frost-burned earth, fading
from brown to gray to blue as they hazed out toward the horizon.
The mountains dipped and heaved, jostled and collided. No towns
or roads were in sight. It occurred to me that I would never be able
to find my way to Maine without the trail. In this foreign, involuted
terrain, I would have struggled to even make it to the next ridge. For
the next five months, the trail would be my lifeline.

+

On a trail, to walk is to follow. Like prostration or apprenticeship, trail
walking both requires and instills a certain measure of humility. To
keep my pack light, I had brought along no maps, no satellite assis-
tance, only a thin guidebook and a cheap compass for emergencies.

The trail was my only real source of navigation. So I clung to it, like Theseus tracing Ariadne's unspooling ball of twine.

In my journal one night I wrote: "There are moments when you cannot help but feel that your life is being controlled by some not-entirely-benevolent god. You skirt down a ridge only to climb it again; you climb a steep peak when there is an obvious route around it; you cross the same stream three times in the course of an hour, for no apparent reason, soaking your feet in the process. You do these things because someone, somewhere, decided that that's where the trail must go."

It was a creepy feeling, knowing that my decisions were not my own. In the first few weeks I often thought back to an anecdote I'd once heard about E. O. Wilson, the famed entomologist. In the late 1950s, to entertain visitors, Wilson used to write his name on a piece of paper with a special chemical liquid. Afterward, a swarm of fire ants would emerge from their nest and dutifully line up to spell out each letter of his name, like members of a marching band.

Wilson's party trick was, in fact, the result of a major scientific breakthrough. For centuries, scientists had suspected that ants left invisible trails for one another, but Wilson was the first to pinpoint the source: a tiny, finger-shaped organ called the Dufour's gland. When he extracted the gland from the abdomen of a fire ant and smeared it across a plate of glass, other fire ants immediately swarmed to it. ("They tumbled over one another in their haste to follow the path I had blazed for them," Wilson recalled.) He later synthesized this trail pheromone, a single gallon of which, he estimated, could summon one trillion fire ants.

In 1968 a group of researchers in Gulfport, Mississippi, put a new twist on Wilson's trick: They discovered that a certain species of termite will even follow a line drawn by a normal ballpoint pen, which contains glycol compounds that termites mistake for trail pheromones. (For some reason, termites prefer blue ink over black.) Ever since, science teachers have amused their students by drawing blue spirals on sheets of paper, while termites line up and confusedly circle toward nowhere.

On my hike, when the trail veered hard to the east or west, I would often wonder whether I too wasn't being led in cruel circles. Seen in a certain light, trails represent a particularly grim form of determinism. "Man may turn which way he please, and undertake any thing whatsoever," wrote Goethe, "he will always return to the path which nature has prescribed for him." On the AT, this was certainly the case. Though I explored the surrounding woods and hitchhiked into towns, in the end I always came back to the trail. If uncertainty is the heart of adventure, I thought to myself, what kind of adventure was this?

+

Northward I moved, through a gray southland spring. The trees were black scrags, the ground papered in old leaves. One morning in Tennessee, I awoke to find my hiking shoes bronzed in ice. In North Carolina, I hiked through knee-deep snow, then ankle-deep slush. The walking was hard, but then every few days, regardless of the terrain or the weather, I would experience the joy of slipping from the dark woods and ascending into the air and light.

In my second week on the trail I fell in with a tight little group of fellow thru-hikers. We happily traveled together for a few weeks. But upon reaching Virginia, I quickened my pace and lost them. Weeks or months later, whenever I slowed down or they sped up, I would bump into these friends again, as if by some miraculous coincidence. The miracle, of course, was the trail itself, which held us together in space like so many beads on a string.

Each of us adopted new trail names. Most people were given their names by fellow thru-hikers because of something they had said or done; my friend Snuggles, for example, had a habit of snuggling up against other hikers in the lean-tos at night to keep herself warm. Others picked names in an attempt to shape new, aspirational identities for themselves. A tense silver-haired woman renamed herself Serenity, while a timid young man called himself Joe Kickass; sure enough, over time, she seemed to grow incrementally calmer, and he more audacious.

A group of jolly older women christened me Spaceman, in reference to the astral appearance of my shiny, ultralight hiking gear. The name clicked. In the trail registers—notebooks located at regular intervals along the trail, meant for recordkeeping and note sharing—I began drawing a series of comic strips. The protagonist was a spaceman who had come down to Earth and somehow found himself navigating the strange customs, odd characters, and pseudo-wildernesses of the Appalachian Trail.

Once a week or so, a group of us thru-hikers would hitchhike into town together, find a cheap motel (sometimes piling six or eight people into a single room), and spend the day showering, washing our filthy clothes, drinking beer, eating impossible quantities of greasy food, and watching bad TV—glutting ourselves, like barbarians, on the meretricious pleasures of civilization. By the next morning we would be eager to get back on the trail, where we could sweat out the gunk and savor the clean air.

I had expected the trail to be a refuge for loners like me; the sense of community that formed among us scattered thru-hikers took me by surprise, and then grew to be one of the hike's nectarine joys. We were bonded by common experience. Each of us knew how it felt to walk for weeks through hail and snow and rain. We starved; we gorged. We drank from waterfalls. In the Grayson Highlands, wild ponies licked the sweat from our legs. In the Smokies, black bears haunted our sleep. We had each faced down the same Cerberus of loneliness, boredom, and self-doubt, and we had learned that the only solution was to out-walk it.

+

As I got to know my fellow thru-hikers—a motley pack of freedom seekers and nature worshipers and outright kooks—it struck me as odd that all of us had willingly confined ourselves to a single path. Most of us saw this hike as an interlude of wild freedom before we reentered the ever-tightening hedge maze of adult life. But complete freedom, it turned out, is not what a trail offers. Quite the opposite—a trail is a tactful reduction of options. The freedom of the trail is riverine, not oceanic.

To put it as simply as possible, a path is a way of making sense of the world. There are infinite ways to cross a landscape; the options are overwhelming, and pitfalls abound. The function of a path is to reduce this teeming chaos into an intelligible line. The ancient prophets and sages—most of whom lived in an era when footpaths provided the primary mode of transport—understood this fact intimately, which is why the foundational texts of nearly every major religion invoke the metaphor of the path. Zoroaster spoke often of the "paths" of enhancement, of enablement, and of enlightenment. The ancient Hindus too prescribed three *margas*, or paths, to attain spiritual liberation. Siddhārtha Gautama preached the *āryāṣṭāṅgamārga*, or the Noble Eightfold Path. The Tao literally means "the path." In Islam, the teachings of Muhammad are called the *sunnah* (again, "the path"). The Bible, too, is crisscrossed with trails: "Ask for the ancient paths, where the good way is, and walk in it, and you shall find rest for your souls," commanded the Lord to the idolaters. (Responded the idolaters: "We will not walk therein.")

There are, it is often said by the more ecumenical prophets, many paths up the mountain. So long as it helps a person navigate the world and seek out what is good, a path, by definition, has value. It is rare to run across a spiritual leader preaching that there are *no* paths to enlightenment. Some of the Zen masters came close, though even the great Dōgen stated that meditation "is the straight path of the Buddha way." The Indian philosopher Jiddu Krishnamurti stands out in this regard. "Truth has no path," he wrote. "All authority of any kind, especially in the field of thought and understanding, is the most destructive, evil thing." Unsurprisingly, his path of pathlessness attracted fewer adherents than the reassuringly detailed instructions of Muhammad or Confucius. Lost in the howling landscapes of life, most people will choose the confinement of a path to the dizzying freedom of an unmarked wilderness.

+

My spiritual path, to the extent that I had one, was the trail itself. I regarded long-distance hiking as an earthy, stripped down, American form of walking meditation. The chief virtue of the trail's confining structure is that it frees the mind up for more contemplative pursuits. The aim of my slapdash trail religion was to move smoothly, to live simply, to draw wisdom from the wild, and to calmly observe the constant flow of phenomena. Needless to say, I mostly failed. Looking back through my journal recently, I found that rather than spending my days in a state of serene observation, much of my time was given over to griping, fantasizing, worrying over logistics, and dreaming of food. Enlightened I was not. But overall I was as happy and healthy as I'd ever been.

Over the course of my first couple of months, my pace gradually increased, from ten miles per day up to fifteen and then twenty. I continued to accelerate as I reached the relatively low-lying ridges of Maryland, Pennsylvania, New Jersey, New York, Connecticut, and Massachusetts. By the time I crossed over into Vermont, I was covering as many as thirty miles a day. In the process, my body was being re-tooled for the task of walking. My stride lengthened. Blisters hardened to calluses. All spare fat, and a fair bit of muscle, was converted into fuel. At any given moment, one or two com-

ponents of the machine were usually begging for maintenance—a
sore ankle, a chafed hip. But on the rare days when everything was
running in harmony, hiking a good stretch of trail felt like gunning
a supercar down an empty interstate: a perfect marriage of instrument and task.

My mind began to change, subtly, too. A legendary old hiker
named Nimblewill Nomad once told me that eighty percent of as-
piring Appalachian Trail thru-hikers who give up do so for mental
reasons, not physical ones. "They just can't deal with the daily, the
weekly, the monthly challenge of being out there in the quiet," he said.
I begrudgingly learned to embrace the monastic silence of the eastern forests. Some days, after many miles, I would slip into a state of
near-perfect mental clarity—serene, crystalline, thought-free. I was,
as the Zen sages say, just walking.

+

The trail leaves its mark upon its travelers: My legs became a map of
black scrapes and leechy pink scars. Ragged holes opened up in my
hiking shoes, and beneath those, in my socks, and beneath those, in
my feet. My T-shirt began to dissolve from the months of friction and
corrosive sweat. If I reached back, I could feel my shoulder blades
pushing through the threadbare fabric like budding wings.

At the same time, I began to notice that we hikers likewise alter
the trail in our passing. I first recognized our impact when climbing
the steep S-shaped turns up hillsides called switchbacks. When a trail
is too curvy, descending hikers tend to create shortcuts to skip the
turns. I also noticed that in boggy areas, hikers would scramble for
dry footing, which split the trail into multiple strands. There seemed
to be a basic conflict between the rationale of the trail's architects and
that of its walkers. Later, by volunteering on trail-building crews, I
would learn why this is so: hikers typically seek the path of least re-

sistance across the landscape. The trail designers, meanwhile, attempt to build trails that will resist erosion, spare sensitive plant life, and avoid private property lines. (The push to teach hikers "Leave No Trace" principles over the past twenty years has had some success in realigning these divergent value systems.) But even if one assiduously stayed within the trail bed, one would still be altering the trail, because every step a hiker takes is a vote for the continued existence of a trail. If everyone decided to stop hiking the AT forever, it would become overgrown and eventually disappear.

Here is where the notion of the spiritual path, as portrayed in countless holy books, falters: scriptures tend to present the image of an unchanging route to wisdom, handed down from on high. But paths, like religions, are seldom fixed. They continually change— widen or narrow, schism or merge—depending on how, or whether, their followers elect to use them. Both the religious path and the hiking path are, as Taoists say, made in the walking.

Use creates trails. Long-lasting trails, then, must be *of use*. They persist because they connect one node of desire to another: a lean-to to a freshwater spring, a house to a well, a village to a grove. Because they both express and fulfill the collective desire, they exist as long as the desire does; once the desire fades, they fade too.

In the 1980s, a professor of urban design at the University of Stuttgart named Klaus Humpert began studying a series of dirt footpaths that had sprung up on the campus's greens, forming shortcuts between paved walkways. He performed an experiment where he erased the campus's informal footpaths by resodding them with grass. Just as he suspected, new trails soon appeared exactly where the old ones had been.

These impromptu trails, which are surprisingly common, are called "desire lines." They can be found in the parks of every major city on earth, slicing off the right angles that efficiency deplores. Studying satellite imagery, I have found desire lines even in the capitals of the world's most repressive countries—in Pyongyang, in Naypyidaw, in Ashgabat. Understandably, dictatorial architects, like actual dictators, despise them. A shortcut is a kind of geographic graffiti, pointing out the authoritarian failure to predict our needs and police our desires. In response, planners sometimes attempt to impede desire lines by force. But this tactic is doomed to failure—hedges will be trampled, signs uprooted, fences felled. Wise designers sculpt *with* desire, not against it.

Previously, when I found an unmarked trail in the woods or across a city park, I used to wonder about its authorship. But usually, I've learned, the answer is that no one person made it. Instead, it *emerged*. Someone made a stab at a problem, took a tentative trip, and the next person followed, and then another, subtly improving the route along the way.

Trails are not unique in this regard—a similar evolutionary process takes place with other communal creations, like folktales, work songs, jokes, and memes. Upon hearing an old joke, I used to wonder what nameless, forgotten comic genius had written it. But this was a futile question to ask, because most old jokes are not born whole; they evolve over the course of decades. Richard Raskin, a scholar of Jewish humor, has sifted through hundreds of anthologies of Jewish jokes in multiple languages, from as far back as the early nineteenth century to the present, to find the origins of classic jokes. What he discovered was that traditional Jewish jokes evolve along common "pathways"—which usually involve reframing, tweaking logic, swapping out characters and settings, and adding more surprising punch lines—all in search of "a better way of fulfilling the stories' comic potential." Like a good trail, a good joke is the result of an untold number of nameless authors and editors. He provides an example from 1928, in which a husband and wife are walking down a dirt road when a heavy rain begins to fall:

"Sarah, pull your skirt up higher. It's practically dragging in the mud!" cries the husband.

"I can't do that. My stockings are torn!" replies his wife.

"Why didn't you put a fresh pair of stockings on?" the husband asks.

"Could I know it was going to rain?"

Raskin deems this joke a failure; it lacks the logical contradiction that lies at the heart of the absurd. But it was a start. Twenty years later, the joke had been tweaked in a number of ways: the setting was moved from an unnamed location to the mythic town of Chelm, which was known to be full of fools; the sentences were sharpened; and the stockings were swapped out for an umbrella, giving the punch line a neater logical paradox. Having passed through countless mouths, the joke had grown from a clunker to a classic:

Two sages of Chelm went out for a walk. One carried an umbrella, the other didn't. Suddenly, it began to rain.

"Open your umbrella, quick!" suggested the one without an umbrella.

"It won't help," answered the other.

"What do you mean, it won't help? It will protect us from the rain."

"It's no use, the umbrella is as full of holes as a sieve."

"Then why did you take it along in the first place?"

"I didn't think it would rain!"

+

One torrential afternoon on the AT, as I was hiking around Nuclear Lake, in New York, I turned a corner to discover a black bear waddling down the middle of the trail. It apparently could neither hear nor smell me amid the rain. It went on calmly snuffling along until I clacked my trekking poles together, at which point it spun around, spotted me, and then nervously trundled off into the woods. I stopped to inspect the stubby-fingered, sharp-clawed prints it had left in the mud. Over the following weeks I began to notice other prints—mostly deer, squirrel, raccoon, and, farther north, moose—pressed into the wet trail. When I left the trail to explore the nearby woods, I was surprised to find a shadow kingdom of trails connecting parts unknown.

Humans are neither the earth's original nor its foremost trail-blazers. Compared to our clumsy dirt paths, the trails of ants are downright wizardly. Many species of mammals, it turns out, are also remarkably adept trail-builders. Even the dumbest animals are experts at finding the most efficient route across a landscape. Our languages have grown to reflect this fact: In Japan, desire lines are called *kemonomichi*, or beast trails. In France, they call them *chemin de l'âne*, or donkey paths. In Holland, they say *Olifantenpad*, elephant paths. In America and England, people sometimes dub them "cow paths."

"We say the cows laid out Boston," wrote Emerson, in reference to the (probably apocryphal) belief that the city's crooked grid was the result of paving old cow paths. "Well, there are worse surveyors. Every pedestrian in our pastures has frequent occasion to thank the cows for cutting the best path through the thicket, and over the hills: and travelers and Indians know the value of a buffalo-trail, which is sure to be the easiest possible pass through the ridge." More than a hundred years later, a study from the University of Oregon has lent credence to Emerson's claim: forty cattle were pitted against a sophisticated computer program and tasked to find the most efficient path across a field. In the end, the cows outperformed the computer by more than ten percent.

Before colonization, many North American tribes followed deer and bison trails, which found the lowest passes across mountain ranges and the shallowest fords across rivers. Elephants, too, are thought to have cleared the most expedient roads through many parts of India and Africa. Nonhuman animals achieve this efficient design not through superhuman intelligence, but through sheer persistence. They continually search for better routes, and once one is found, they adopt it. In this manner, trail networks of incredible efficiency can arise simply, organically, iteratively, without any forethought necessary.

A clever and patient observer can watch a trail sleeken in real time. The physicist Richard Feynman, for instance, witnessed this phenomenon while studying the ants that infested his home in Pasadena. One afternoon, he took note of a line of ants walking around the rim of his bathtub. Though myrmecology was far from his area of expertise, he was curious to find out why ant trails inevitably "look so straight and nice." First, he placed a lump of sugar on the far side of the bathtub and waited for hours until an ant found it. Then, as the ant carted a piece of the sugar back to its nest, Feynman picked up a colored pencil and traced the ant's return path along the bathtub. The resulting trail was "quite wiggly," full of errors.

Another ant emerged, followed the first ant's trail, and located the sugar. As it plodded back to the nest, Feynman marked its trail with a different color of pencil. But in its haste to return with its bounty, the second ant repeatedly lost the first ant's trail, cutting off many of the unnecessary curves: The second line was noticeably straighter than the first. The third line, Feynman noted, was even straighter than the second. He ultimately followed as many as ten ants with his pencils, and, as he'd expected, the last few trails he traced formed a neat line along the bathtub's edge. "It's something like sketching," he observed. "You draw a lousy line at first; then you go over it a few times and it makes a nice line after a while."

I later learned that this streamlining process extended beyond ants, or even animals. "All things optimize in nature, to some degree," an entomologist named James Danoff-Burg told me.

Intrigued, I asked him if there was a good book I could read on optimization.

"Sure," he said. "It's called *The Origin of Species* by Charles Darwin."

Evolution, he explained, is a form of long-term, genetic optimization; the same process of trial and error takes place. And, as Darwin showed, in the great universal act of streamlining, even the errors are essential. If some ants weren't error-prone, the ant trail would never straighten out. The scouts may be the genius architects who blaze the trails, but any rogue worker can be the one who stumbles upon a shortcut. Everyone optimizes, whether we are pioneering or perpetuating, making rules or breaking them, succeeding or screwing up.

+

After three and a half months I reached the base of Mount Washington in New Hampshire. I climbed it via the Crawford Path, the same trail I had hiked when I was ten. In rapid succession I pieced together a half-dozen peaks that I'd climbed at different times in the past decade: the Presidentials, Old Speck, Sugarloaf, Baldpate, the Bigelows. The order of the mountains sometimes surprised me; it was as if someone had opened my childhood photo album and rearranged my memories. The mountains also seemed smaller than I remembered. Hikes that had taken days when I was a kid now took only hours. It was an eerie sensation—that same uncanny, gargantuan feeling you get from revisiting your old kindergarten.

Any feeling of mastery I harbored was mingled with feelings of humility. I had hiked two thousand miles, but I could never have gotten there on my own. My route had been carved out by scores of volunteer trail-builders and a continuous flow of prior walkers.

I often felt this way on the trail: I was able to hold both one notion and its direct opposite in my mind at the same time. Paths, in their very structure, foster this way of thinking. They blear the divide between wilderness and civilization, leaders and followers, self and other, old and new, natural and artificial. It is fitting that in Mahayana Buddhism, the image of the Middle *Path*—and not some other metaphor—is used as a symbol of dissolving all dualities. The only binary that ultimately matters to a trail is the one between use and disuse—the continual, communal process of making sense, and the slow entropic process by which it is unmade.

+

On August 15, almost five months to the day after I had started out from Springer Mountain, I reached the summit of Mount Katahdin in Maine. Far below, in every direction, were green forests and blue lakes and islands of green forest within the blue of the lakes. After what felt like months of steady rain, the skies had finally cleared. I could feel the dampness baking from my bones. I had at last reached the trail's end.

In the center of the peak was an iconic wooden sign announcing the trail's northern terminus. It had the air of a shrine. Groups of day hikers hung back from it, forming a respectful half-circle, while a handful of thru-hikers approached it, one by one, with looks of reverence and tamped expectation. Each hiker had a moment alone with the sign, posed for a photograph to commemorate the occasion—some exuberant, some somber—and then moved on, allowing the next hiker to approach.

When my turn came, I walked up to the sign, laid my hands on it, and kissed its wind-scoured surface. The moment held a certain surreal quality; I already had imagined it a thousand times. My friends and I popped a bottle of cheap champagne, which we shook and sprayed in fanning arcs into the air. When we finally took a sip, the champagne had already gone flat and warm. That was a rough analog for how it felt to finish the trail: buzzy and yet weirdly dull. After five months, it was over.

And yet, when I moved back to New York City, I found that I continued to look at the world with the eyes of a thru-hiker. After almost half a year spent in mountainous wilderness, the city seemed at once a marvel and a monstrosity. It was hard to imagine a space more thoroughly transformed by human hands. What struck me most, though, was its rigidity: straight lines, right angles, cement roads, concrete walls, steel beams, harsh rules regulated by force. Waste was rampant; everything broke. The trail had taught me that good designs—like age-old tools and classic folktales—are *trail-wise*: They fulfill a common need by balancing efficiency, flexibility, and durability. They streamline. They self-reinforce. They bend but do not break. So much of our built environment, by comparison, seemed terribly, perilously inelegant.

Meanwhile, everywhere I looked, I noticed new trails: a desire line winding through a tiny park beside the East River, a line of ants inching along my windowsill. I noticed how the shoes of passing commuters wore greasy lines into the concrete of the subway platform, how spots of blackened chewing gum and flattened cigarette butts marked the entrance to nightclubs. Reading omnivorously, I discovered trails running thick through works of literature, history, ecology, biology, psychology, and philosophy. Then I put the books down and walked some more, seeking out fellow travelers—trail-walkers and trail-builders, hunters and herders, entomologists and ichnologists, geologists and geographers, historians and systems theorists—in the hopes of gleaning some common truths from their diverse fields of expertise.

Somewhere along the way I realized that at the heart of my thinking lay a simple idea: A trail sleekens to its end. An explorer finds a worthwhile destination; then every walker who follows that trail makes it a little better. Ant trails, game paths, ancient ways, modern hiking trails—they all continually adapt to the aims of their walkers. Hurried walkers make straighter paths and leisurely walkers make curvier ones, just as some societies seek to maximize profit, while others strive to maximize equality, or military might, or gross national happiness.

The path of a runner often diverges from that of a walker, because, though both may be headed to the same place, they do so with differing priorities. A New Zealand sheep farmer named William Herbert Guthrie-Smith once observed that horse trails in open country will gradually straighten out. However, he noticed that this only took place in areas where the horses were allowed to trot, canter, or gallop. At a slow walk, the horses gladly followed each turn of a sinuous trail, minimizing their work by bending with the contours of the topography. When they sped up, they began to cut the inside corners off the curves, straightening them. If the horses had been allowed to run "at racing speed," Guthrie-Smith believed they would "in time rule out paths almost perfectly straight."

The lesson to be found here is not just that the trail of a galloping horse streamlines. It is that both the fast horse and the slow one seek the path of least resistance. When aims differ, trails do too. These overlapping and crisscrossing trails, created by countless living beings pursuing their own ends, form the planet's warp and woof.

+

This book is the culmination of many years of research and many miles of walking. Throughout, I was fortunate to have been guided by experts in their fields, each illuminating a key element in the long history of trails, spanning from the Precambrian to the postmodern. In the first chapter, we take a close look at the world's oldest fossil trails, and explore the question of why animals first began to move. The second chapter investigates how insect colonies create trail networks to maximize their collective intelligence. In the third chapter, we follow the trails of four-legged mammals like elephants, sheep, deer, and gazelles, to learn how they manage to navigate immense territories, and how our efforts to hunt, herd, and study them have shaped our development as a species. Chapter four chronicles how ancient human societies stitched together their landscapes with networks of footpaths, which then became tightly interwoven with the vital cultural threads of language, lore, and memory. In the fifth chapter, we unearth the winding origins of the Appalachian Trail, and other modern hiking trails like it, which stretch back centuries to Europeans' colonization of the Americas. In the sixth and final chapter, we trace the longest hiking trail in the world from Maine to Morocco, and we discuss how trails and technology—having combined to create our modern transportation system and communication network—connect us in previously unimaginable ways.

As a writer and a walker, I am limited by my experience, my background, and my place in history. If this book strikes some readers as too Americentric, or too anthropocentric, I beg their forgiveness; I am, after all, just one American human, doing my best to make sense of a deceptively complex topic. It is also important to note that although the structure of this book is loosely spatial and chronological—moving from the tiny and ancient to the huge and futuristic—this book is not what philosophers call a teleology, a succession of rungs leading up to an ultimate goal. I am not so foolish as to believe that trails have been evolving for hundreds of millions of years only to culminate in the hiking paths of the twenty-first century. I urge readers to avoid interpreting this book's structure as a ladder leading upward, but to instead regard it as a trail winding from the dim horizon of the past to the wide foreground of our present circumstances. Our history is one of many paths we might have taken, but it was the one we took.

Trails can be found in virtually every part of this vast, strange, mercurial, partly tamed, but still shockingly wild world of ours. Throughout the history of life on Earth, we have created pathways to guide our journeys, transmit messages, refine complexity, and preserve wisdom. At the same time, trails have shaped our bodies, sculpted our landscapes, and transformed our cultures. In the maze of the modern world, the wisdom of trails is as essential as ever, and with the growth of ever-more labyrinthine technological networks, it will only become more so. To deftly navigate this world, we will need to understand how we make trails, and how trails make us.

CHAPTER 1

IT IS IMPOSSIBLE to fully appreciate the value of a trail until you have been forced to walk through the wilderness without one. There is a practical reason why, for more than a thousand years, after the fall of Rome and before the rise of Romanticism, little was more abhorrent to the European mind than the prospect of a "pathless" or "tangled" wilderness. Dante famously described the feeling of finding oneself in a "wild, harsh and impenetrable" forest without a path as "scarcely less bitter than death."

Five hundred years later, a Romantic like Lord Byron could proclaim that there is "a pleasure in the pathless woods," but only once the wilds of Western Europe had been tamed and caged. By that point, the true "pathless wilderness" was believed to exist only on other continents, like North America, where the phrase was still being used well into the nineteenth century.* The American wilder-

---

*This, despite the fact that the land had been webbed with native footpaths since long before white people arrived.

ness came to symbolize an inhospitable and far-off land, cold, cruel, and uncivilized. At the Boston Railroad Jubilee in 1851, the politician Edward Everett described the land between Boston and Canada as a "horrible wilderness, rivers and lakes unspanned by human art, pathless swamps, dismal forests that it made the flesh creep to enter . . ."

Pathless wildernesses still exist in the modern world, and at least some have retained their power to elicit dread. I have visited one such place. It lay on the northern rim of a glacial fjord called Western Brook Pond, on the island of Newfoundland, in Canada's easternmost province. If you want to be taught (however harshly) the blessing of a well-marked trail, go there.

To cross the fjord's stygian waters, I had to hire a ferryboat. Aboard the ferry, the captain explained that the water below the boat was so pure (in a hydrologist's terms, so ultraoligotrophic) that it bordered on nonexistence; he said it played havoc with the sensors in modern water pumps because the water couldn't even carry an electrical current.

On the far side of the fjord, the captain dropped me and four other hikers off at the base of a long ravine, where a series of animal trails led through a dense fern jungle and up a granite cliff face bisected by a waterfall. This was my first hiking trip since returning home from the Appalachian Trail. I felt strong; my pack was light. Weaving through the tall ferns, I quickly passed the other hikers. At the top of the ravine I found a vast green tableland. The trail I had been following vanished altogether. Soaked in sweat from the hike up, I took a moment to rest, my feet dangling over the cliff's edge. At the ragged western edge of the tableland, it abruptly dropped hundreds of feet to the fjord's indigo water.

I sat and watched as the other hikers wound their way up the ravine. Once they had reached the top, the other four hikers all headed south, along a more scenic route. Watching them go, laboring beneath their heavy packs, I felt a swell of confidence. I rose, map and compass in hand, and headed north. *This shouldn't be too tough,* I thought. *After all, it's only sixteen miles.*

As I began hiking, however, that confidence soon withered. One might suppose that, after a lifetime of walking within the rigid confines of trails and pathways—from wilderness footpaths to the moving walkways in airports—it would come as a relief to roam free in any direction. But this was not the case. A low bass beat of terror throbbed behind my every decision. I was alone, and without any means of communication save a small, park-issued radio locator beacon, which resembled a large plastic pill with a wire hanging out of it. It could be used, I had been assured, to track me down if the park ranger's office hadn't heard from me for more than twenty-four hours after my scheduled return. It seemed a wonderful device for recovering corpses.

More bedeviling though, was the sheer number of minuscule choices I was forced to make at each turn. Even with a rough idea of where I was meant to go, there were still countless decisions to make at any one moment: whether to slant uphill or down; whether this tuft of grass or that would support my weight as I tiptoed across a bog; whether to hop along the rocks on a lakeshore's edge or bash my way through the bush. In every landscape, as in every mathematical proof, there are countless routes one can take to the solution, but some are elegant and others are not.

My navigational woes were compounded tenfold by the problem of what Newfoundlanders call "Tuckamore"—groves of spruce and fir that have been dwarfed by strong winds. From afar, the trees resemble a scrum of fairy-tale hags, all hunches and claws. Like most elfinwood trees, they can grow for centuries without ever reaching any higher than one's chin. What they lack in height, they gain in hardiness.

Countless times on my hike, I would reach a section where a small grove of Tuckamore stood between me and where I needed to be. I would glance at my watch to mark the time, estimating it should take no more than ten minutes to cross. Then I would take a deep breath and enter the low green copse. It was like dipping into a nightmare. Suddenly the air was dark, and the space apportioned chaotically. As I fought to take each step, branches clawed red gashes into my skin and pulled the water bottles from the pockets of my backpack. Out of frustration, I tried stomping on the trees, to break them, or at least to punish them, but to no avail; they sprang back, unharmed. Here and there a set of moose or caribou prints would form a narrow, muddy game trail, but after a short while it would dwindle or veer astray. Off to the left, a pocket of sunlight would appear, and I would follow it, only to find a pool of mud. It was like moving through a labyrinth that left you no choice but to, from time to time, lower your shoulder and charge your way *through the walls.*

At last, exhausted and bleeding, I would emerge. My watch would reveal that an hour had passed, and I had covered no more than fifty yards.

Eventually, I learned to pick my way through these mazes by watching the movements of the moose. One trick moose use is to follow waterways, which, though muddy, often find the most expedient path through a thicket. They also walk with high, arching steps to flatten the branches underfoot. It was in perfecting this technique that I came to my greatest revelation: at one point near the end of the hike, I found that by counterintuitively selecting the densest bunches of Tuckamore, I could actually lift myself up and walk along the tops of the trees like a *wuxia* warrior.

By nightfall on the second day, I was at least two miles off course. It had already taken me a day longer than I had expected to hike the mere sixteen miles, and not once had I spent the night on level ground or near fresh water.

All night a light rain fell. Around dawn, I awoke from my bivouac high atop a ridge to observe a wide band of hyacinth sky moving toward me. At first I perceived this lovely sight as a break in the clouds and lay back down to sleep. But as I turned back to my sleeping bag, I noticed the purple stripe was finely veined with lightning. It was not clear sky, I realized, but a massive storm cloud stretching from one end of the horizon to the other. It let out a soft digestive growl.

Within the space of a half hour, the storm cloud rushed overhead. The air was crazed with rain. Fearing a lightning strike, I scrambled out of my sleeping bag, out from under my tarp, and down to the lowest point I could find. There I crouched on my sleeping mat on the balls of my feet, hands over my head, shaking and drenched, as delicate strings of light detonated all around me.

For the better part of an hour, awash in mounting waves of tympanic rumble, I had time to reconsider the merits of hiking. Stripped of its Romantic finery, the wild ceased to inspire; only a gauzy scrim separated sublimity and horror. Jacques Cartier, upon visiting this island in 1534, declared that he was "inclined to believe that this is the land God gave to Cain." He was right. It was a dark and pestilential place. The apparent beauty was only a ruse to lure you into its flytrap maw. I vowed to myself that if I made it out of this alive, I would never hike again.

Upon seeing the Earth's true brutality unmasked, authors throughout history have expressed a similar sense of disillusionment, even betrayal. In his semiautobiographical short story "The Open Boat," Stephen Crane captured the chilling moment when a shipwreck victim realizes that nature is "indifferent, flatly indifferent." Annie Dillard—after watching a giant water bug gruesomely devour a frog—grapples with the possibility that "the universe that suckled us is a monster that does not care if we live or die." Goethe went one step further, calling the universe "a fearful monster, forever devouring its own offspring." Kant, Nietzsche, and Thoreau all describe nature not as a mother, but as a "stepmother"—a winking reference to the wicked villainesses of German lore.

The English writer Aldous Huxley came to this realization while walking through the wilds of Borneo. Being fussy about his lodgings and terrified of cannibals, Huxley preferred to stick to "the Beaten Track." But one day eleven miles outside of Sandakan, the paved road he was traveling along abruptly ended, and he was forced to trek through the jungle. "The inside of Jonah's whale could scarcely have been hotter, darker or damper," he wrote. Lost in that mute, hot twilight, even the cries of birds startled him, which he imagined to be the whistles of devilish natives. "It was with a feeling of the profoundest relief that I emerged again from the green gullet of the jungle and climbed into the waiting car. . . . I thanked God for steam-rollers and Henry Ford."

Back home, Huxley drew from this experience to compose a series of audacious attacks against the Romantic love of wilderness. The worship of nature, he wrote, is "a modern, artificial, and somewhat precarious invention of refined minds." Byron and Wordsworth could only rhapsodize about their love of nature because the English countryside had already been "enslaved to man." In the tropics, he observed, where forests dripped with venom and vines, Romantic poets were notably absent. Tropical peoples knew something Englishmen didn't. "Nature," Huxley wrote, "is always alien and inhuman, and occasionally diabolic." And he meant *always*: Even in the gentle woods of Westermain, the Romantics were naive in assuming that the environment was humane, that it would not callously snuff out their lives with a bolt of lightning or a sudden cold snap. After three days amid the Tuckamore, I was inclined to agree.

Once the rain had ceased, I shook the water from my tarp, packed my things, and began walking to get warm. I found myself looking with new admiration upon the Tuckamore, which looked unfazed by the storm—nourished, even. Those rugged little trees were perfectly fitted to their niche, sculpted by the wind, deeply rooted to their land. I, meanwhile, was a perpetual wanderer, ill-equipped, maladapted, and lost.

Three hours later, after a few more harrowing misadventures (ravines descended in vain; waterfalls tenuously traversed), I found my

way to the endpoint of the unmarked wilderness, where a large pyr-
amidal pile of rocks marked the beginning of the trail back down to
Snug Harbour. I whooped and hollered, awash in the same relief Huxley felt upon spotting his chauffeur. The trail, however rough, would
return me to the human realm. Delivered from chaos, I promptly
forgot my former terror, fell in love with the earth anew, and once
again desired to walk every inch of it.

+

I had not traveled to Newfoundland to be mauled by trees. The hike
was a mere diversion, a side trip. My ultimate destination was a yet
more baffling and inaccessible wilderness: the distant past. I was making my way to a rocky outcropping on the island's southeast corner,
where I hoped to find the oldest trails on earth.

These fossil trails, which are roughly 565 million years old, date
back to the dimmest dawn of animal life. Now fossilized and faint,
each one is roughly a centimeter wide, like a fingertip's errant brush
across the surface of a drying clay pot. I had read all about them,
but I wanted to touch them, to trace their runnels like a blind man.
I hoped that encountering them up close would resolve a question
I've long harbored, like an old thorn: Why do we, as animals, uproot
ourselves rather than maintaining the stately fixity of trees? Why do
we venture into places where we were not born and do not belong?
Why do we press forward into the unknown?

+

The world's oldest trails were discovered one afternoon in 2008 by
an Oxford researcher named Alex Liu. He and his research assistant
were scouting for new fossil sites out on a rocky promontory called
Mistaken Point, where a series of well-known fossil beds overlook
the North Atlantic. Bordering one surface, Liu noticed, was a small
shelf of mudstone that bore a red patina. The red was rust—an

oxidized form of iron pyrite, which commonly appears in local Precambrian fossil beds. They scrambled down the bluff to inspect it. There, Liu spotted what many other paleontologists before him had somehow missed: a series of sinuous traces thought to be left behind by organisms of the Ediacaran biota, the planet's earliest known forms of animal life.

The ancient Ediacarans, which likely went extinct around 541 million years ago, were exceedingly odd creatures. Soft-bodied and largely immobile, mouth-less and anus-less, some were shaped like discs, others like quilted mattresses, others like fronds. One unfortunate type is often described as looking like a bag of mud.

We can envision them only dimly. Paleontologists don't know what color they were, how long they lived, what they ate, or how they reproduced. We do not know why they began to crawl—perhaps they were hunting for food, fleeing a mysterious predator, or doing something else entirely. Despite all these uncertainties, what Liu's trails undoubtedly suggest is that 565 million years ago, a living thing did something virtually unprecedented on this planet—it shivered, swelled, reached forth, scrunched up, and in doing so, at an imperceptibly slow pace, began to move across the sea floor, leaving a trail behind it.

+

To reach the fossil trails at Mistaken Point, I flew to the town of Deer Lake and hitchhiked some seven hundred miles, taking a slow circuitous route that touched almost every corner of the island. Along the way I hiked mountains, swam in rivers, tasted icebergs, camped out, and slept on strangers' couches. Newfoundland is ideal for bumming around; it has one of the lowest homicide rates in the world, the people are generally congenial, and everyone seems to own a big automobile. Car ride by car ride, I made my way down to the island's southeastern tip.

However, when I finally arrived at the park entrance, I was turned away. A vigilant park ranger forbid me to see the trails because I had failed to acquire the proper permits. Their location, I learned, was a matter of great secrecy due to the rise of so-called "paleo-pirates," who had been known to carve out the more notable fossils and sell them to collectors.

Undeterred, I returned the following year—armed, this time, with the proper clearance. A saintly couple I had met the year before graciously offered to pick me up at the airport and give me a ride down to Trepassey, a town nicknamed the "Harbor of the Dead," because its foggy waters had been the site of many shipwrecks. There, at an unprepossessing restaurant in the Trepassey Motel, I finally met with Alex Liu.

Having only read about him in press clippings, I imagined Liu as I did all paleontologists: gray at the temples, a pair of Savile Row spectacles perched on his nose, and behind them, the deep-creased eyes of a man who spends his days peering at small things lit by a harsh sun. But when Liu appeared in the doorway of the restaurant, I was surprised to discover a fresh-faced, raven-haired young man, not yet thirty, with a shy smile. Beside him were his two research assistants: Joe Stewart, who had the shorn head and handsomely punched-up physiognomy of a rugby player, and Jack Matthews, the youngest member of the group, whom I seemed to have caught in a brief hiatus in his metamorphosis from a mischievous boy into a kooky, brilliant, snowy-haired professor.

We shook hands, sat down, and ordered a round of beer and plates of fried fish. They ate heartily. Because money was tight, the team spent two out of every three nights in tents set up in an abandoned trailer park and the third night here at the motel to shower up and wash their clothes. Journalism, they assured me, was not the only field with dwindling resources. Each year, said Liu, university and government budgets for the dusty science of paleontology grew stingier. He smiled with resignation. "What I do is immensely important for understanding where we came from, but it has little wider social impact," he said. "It's not going to solve climate change. It's not going to boost the economy."

As a boy, Liu had loved dinosaurs, particularly those in *Jurassic Park*. The romance of those craning beasts, which he never fully outgrew, coupled with his love of fieldwork and knack for geology, drew him to fossil hunting. When he was pursuing his master's degree at Oxford, he had planned to study ancient mammals, but he found the field crowded; his thesis project was spent studying the teeth of Eocene-era elephants in Egypt. For his PhD work, he turned to the much older and largely unstudied Ediacarans. "If I had taken on a mammal project, then I'd have been trying to answer questions that people have looked at for hundreds of years," he said. "Whereas I knew that Ediacaran stuff was new, uncertain. And that was more enticing, really, because the questions are bigger."

Of all the manifold questions surrounding these elusive, soft-bodied animals, the biggest of all might concern the origins of animal movement. Some paleontologists theorize that the first Ediacaran trail-maker may have set off a series of morphological changes leading, in fits and starts, from a serene garden of swaying anemone-like creatures to today's violent, skeletonized kingdom of sprinters, jumpers, flyers, swimmers, diggers, and walkers. It is rare in science to run across a big new question, and harder still to answer it, but Liu seemed to have this one by the scruff of its neck.

+

For a respectable scientist, wading into the murky world of the Ediacarans is a treacherous endeavor. Information about that distant era is extremely limited, and even the most basic assumptions often prove unreliable. For instance, we still do not know for certain which kingdom of life the Ediacarans belonged to. At various times, it has been proposed that they could have been plants, fungi, colonies of single-celled organisms, or, according to the trace fossil expert Adolf Seilacher, a "lost kingdom" called Vendobionta. While most Ediacaran researchers tentatively agree that they were animals, recently, some have begun arguing that lumping all the known Ediacaran species into one kingdom or another may be too reductive, and each fossil must instead be re-taxonomized one by one.

Sitting next to him at dinner that night, it seemed odd to me that Liu, a soft-spoken and exceptionally careful researcher, was drawn to such a field. Liu told me he first became interested in Ediacarans during a class in his second year at Oxford with a professor named Martin Brasier, who spoke inspiringly about the mysteries of Precambrian fossils. Brasier—who died in a car accident in 2014, at the age of sixty-seven—was a Shiva-like figure among Ediacaran paleontologists, slashing down flimsy theories and widening the domain of *that which cannot be definitively stated*. In his 2009 book *Darwin's Lost World*, Brasier briskly disassembled the principle of uniformity, which decrees that, natural laws being uniform, fossils can be better understood by studying living animals. Uniformitarianism has proved a powerful tool in many fields, Brasier admitted, but it ignores an organism's profound interdependence with its environment. The theory's efficacy therefore begins to break down in the Precambrian era, when there existed a radically altered oceanic ecosystem. "The world before the Cambrian was, arguably, more like a distant planet," wrote Brasier.

To us land-dwellers, even the present-day deep sea is foreign, a crushing black space haunted by spectral oddities: glass squids, carnivorous jellyfish, a fever dream of fluorescence. But in the time when the Ediacarans thrived, the oceans were stranger still. The first Ediacaran to begin crawling around would have discovered a world devoid of predatory animals, with a seafloor covered either in thick bacterial mats or toxic sediment, and, possibly, a climate thawing from a worldwide glaciation event known as "Snowball Earth" (or, more recently, "Slushball Earth"). If that pioneering Ediacaran could see, it would have discovered an underwater desert patchily carpeted with gelatin. Here and there it may have spotted other, nonmobile Ediacarans, which resembled fleshy leaves, many-tendriled sea anemones, or low, round blobs: a whole world populated by brainless, jelly-quivering do-nothings.

The mystery Liu was trying to unravel—regarding the origins of animal movement—is central to solving the larger mystery of how that alien planet transformed into the natural world we all know. Muscular locomotion could have allowed animals to graze on the beefsteak-like bacterial mats and to attack other stationary organisms. The invention of violence might then have kicked off a biological arms race, prompting organisms to evolve hard shells and sharp teeth, the shields and swords that characterize the Cambrian fossil record. This hardening of animal bodies eventually led to the rise of trilobites and tyrannosaurs and Eocene-era Egyptian elephants—and us.

Before the discovery of Ediacaran fossils, and even for a while afterward, many prominent scientists argued that complex life began at the dawn of the Cambrian era. Looked at from a certain angle, the fossil record seemed to support this theory. Around 530 million years ago, like a symphony warming up, the fossil record began teeming with a cacophony of different fossil types. Further back than that was nothing: silence. Some scientists, like Roderick Murchison, a geologist and devout Christian, believed that this lack of evidence was geologic proof of a biblical genesis. ("And God said, 'Let the water teem with living creatures . . .'")

Charles Darwin cautioned against this interpretation, writing in *On the Origin of Species* that, "We should not forget that only a small portion of the world is known with accuracy." He saw the entire geologic record as a history book stretching across multiple volumes. "Of this history we possess the last volume alone, relating only to two or three countries," he wrote. "Of this volume, only here and there a short chapter has been preserved; and of each page, only here and there a few lines."

The truth, it now seems clear, is that Precambrian animals had existed in great numbers, but, being soft-bodied, had not lent themselves to fossilization. They crop up exceedingly rarely, in places like Mistaken Point, where the geologic conditions were just right.

After our dinner at the Trepassey Motel, once our plates had been taken away and dessert politely declined, Liu mentioned that another big question he has yet to answer is why the Ediacaran fossils of Newfoundland are so unusually well preserved. He suspected the Mistaken Point assemblage was smothered in a Pompeii-like flow of volcanic ash and impressed into the bacterial mats on the seafloor. He would have liked to test this hypothesis in a lab, but it had proved tricky, because he would need fresh volcanic ash.

Fortunately, Liu's girlfriend, Emma, was a volcanologist.

"Have you got Emma running around with a bucket to collect you some ash?" Stewart asked, grinning.

"I've asked her if she would," Liu nodded, sincerely. "She was in Montserrat, in the Caribbean, last summer, and that's exactly the right type of ash. But it didn't erupt."

Stewart laughed. "You may be the only man on the planet," he said, "who, when his girlfriend goes to the Caribbean, hopes the volcanoes will erupt."

+

Around our second round of beers, the scientists' conversation turned to the topic of humans. They noted that research into the origins of life provokes an irrational vitriol in many people. Liu mentioned that one of his supervisors, upon publishing a paper about a fifty-million-year-old monkey fossil he had discovered, soon began receiving death threats from creationists. I was reminded of a similar story I'd heard from a former tour guide in New Hampshire. During one of her bus tours, she had mentioned to a group of children that the granite cliffs visible through their windows were some two hundred million years old. The students' chaperone jumped up and wrenched the microphone from the guide's hand to assure the children that what she'd *meant* to say was that the rocks were two *thousand* years old. Covering the microphone, the chaperone explained to the tour guide that it was their church's teaching that the universe was created by God only six thousand years ago. She asked the tour guide to, in the future, please be a bit more respectful of people's differing belief systems.

Liu wryly remarked that he would have little trouble disproving such an assertion.

"But you can't," Stewart said. "Because whatever evidence you put in front of them, they're going to say it's the devil deceiving you."

These words pinged around in my head as I bid them goodnight and started off down the darkened road to the town's beach, where I planned to camp for the night. A deceitful demon: the very same one Descartes summoned in 1641. How, the great cogitator had asked, do we know that what we see is not a pure hallucination, perpetrated upon us by a malignant, godlike figure? How do we know that what we perceive is really the world?

Aldous Huxley, having never forgotten the horror of his "stroll in the belly of the vegetable monster" in Borneo, went on to expand his prickly view of the wilderness into a kind of broad Kantian skepticism about the capacity of humans to ever directly experience reality. He cast the world-in-itself as a place of "labyrinthine flux and complexity," which we are able to navigate only through imagination and invention. "The human mind cannot deal with the universe directly," he wrote, "nor even with its own immediate intuitions of the universe. Whenever it is a question of thinking about the world or of practically modifying it, men can only work on a symbolic plan of the universe, only a simplified, two-dimensional map of things abstracted by the mind out of the complex and multifarious reality of immediate intuition."

Huxley believed that knowledge, even when empirically proven, is only ever a map, never a view of the territory itself. But perhaps it is not so stark as that: perhaps knowledge is more like a trail—a hybrid of map and territory, artifice and nature—wending through a vast landscape. While science may provide a more reliable route to certain answers than, say, a creation myth, it remains narrow; it can reduce the environment to a navigable line, but it cannot encompass it. To a fervent believer in the scientific method, this thought can be unsettling. Great mysteries surround us all, like beasts slinking silently through the night—their presence can be intuited, or imagined, but never fully illuminated.

Paranoia blew gently on my neck as I combed the beach for a suitable place to set up my tent that night. I became convinced that wherever I chose to sleep, local troublemakers would decide to harass me during the night. I feared that in the town's eyes I was seen as a homeless person, a foreign body to be expunged.

I erected the tent on a flat spot close to the road, but the headlights of each passing car swept over the tent, setting it aglow like a paper lantern. I could hear the cars' passengers speaking in parabolas of intelligibility as they bent past. A few remarked on the oddity of my impromptu campsite, so I picked up the tent and moved it farther down the beach, where it was darker. In the long headlights of those cars, my shadow resembled that of a giant carrying an igloo.

At first I selected the flattest spot I could find, but I realized that I was squarely in the path of a set of 4x4 tracks coming from a nearby house. Later in the night, I would hear drunk teenagers speeding down that same path where I might have slept. Beer bottles tinkled onto the sand. At least one rider, a girl, spotted my encampment and said, "Oh, weird, there's a tent down there." I envisioned these antibodies gathering unseen around the tent, smiling, fingers to their lips.

As I lay awake, listening for the faint crunch of approaching foot-steps, I thought back to something mentioned by one of the drivers I'd met while hitchhiking down to Mistaken Point. As she drove south along the coast, she had pointed to the hills to the west and told me that, not long ago, the countryside of Newfoundland was believed to be crisscrossed with "fairy paths." Even now, she said, people occasionally reported seeing small blobs of light floating down these trails.

A fear of fairies traditionally prevented Newfoundlanders from building their houses over old paths. According to Barbara Gaye Rieti's exhaustive folk history *Newfoundland Fairy Traditions*, those who obstructed fairy paths often heard strange sounds in the night, which, in at least one documented case, induced a nervous breakdown. Worse horrors still were visited upon their children; parents would return from some chore to find their baby missing, or lying paralyzed in its crib, or sitting open-mouthed with pain, its head grotesquely enlarged. Sometimes, instead of a baby, they would find a very small, very old person sitting upright in the bassinet, its hair whitened and its fingernails grown long and curled. In one especially nasty tale, a girl in St. John's made the mistake of walking across a lane that ghosts frequented at night. As she crossed, she felt something smack the side of her head, which left a bruise. Back home, the bruise worsened and became infected. "A few days later," Rieti wrote, "the infection broke and pieces of old cloth, rusty nails, needles, and bits of rock and clay were all taken from her face."

As we had cruised south, the driver recounted stories of her family's encounters with ghosts, fairies, white ladies, goblins, gypsies, and angels. She described in detail a time when a ghost or an angel—she and her husband quibbled over which it was—enveloped her in its arms and prevented her from being struck by a car while she was walking down a snowy road at night. Afterward, she sensed that the angel was following her home. When her dog rushed out of the house to greet her, it trotted right past her and stood at the end of

the driveway with its snout angled upward, as if it were being petted
by an invisible hand.

These stories unnerved me, because many of the details were so
utterly mundane. The world looks clear and rigid in the bright light
of the metropolis, but out here on the edge of the continent, in the
murky night and gray fog, anything seemed possible.

+

I awoke to a glassine dawn. Overnight the wind had gusted so hard
it had ripped out two of the tent stakes. The beach was empty, blown
clean. I groggily squirmed out of my sleeping bag, flattened the tent,
and packed my things.

The night before I had agreed to meet Liu's team at the motel
for an early breakfast so we could spend the day fossil hunting. After
breakfast, Stewart and I pillaged the local grocery store for picnic
supplies—white bread, industrial chocolate chip cookies, hickory-
flavored potato sticks, and icy plums ("to keep away the scurvy," he
joked)—and then piled into the research team's rental car, a Japanese
SUV. The synthetic interior bore that rental car smell, the odor not
of something new, but of something smudgily erased. The cargo area
was packed with climbing ropes, a coil of metal wire, a yellow hard
hat, blue aluminum camping bowls, a huge bag of Doritos, sleeping
bags, a tent pole held together with electrical tape, a rock hammer,
an inflatable raft, and tubs of platinum silicone rubber called Dragon
Skin, which was used to make flexible casts of the fossil beds. If only
its next renter could know what strange sights that vehicle had seen.

Liu's plan for the day was to begin our tour at a prominent fossil
site called Pigeon Cove, and then work our way forward in time,
covering about ten miles on foot and by car. We would visit each
of the area's most impressive fossil beds, culminating at the surface
where Liu had discovered the fossil trails.

Windows open to the hard sea wind, we raced across a landscape of stooped trees and yellowing grass to Pigeon Cove, where we got out and hiked down a dirt path to the seaside. There lay a flat slab of rock, the size and texture of three cracked concrete tennis courts, which sloped down into the sea. Its surface was a swirl of gray, chalkboard green, and dusty eggplant. Impressed into it were faint but distinct symbols. One looked like a fleshy frond. Another looked like an arrowhead, but in life probably resembled one of those conical corn snacks sold at gas stations, with its narrow end stuck into the ground. A third, which paleontologists call a "pizza disc," was just a big, bubbly mess.

The team split up and set to their work. Liu pulled out a small black notebook and began making notes about the fossil surface in a neat semi-cursive, complete with illustrations and GPS coordinates. Stewart got down on his knees and began using a clinometer to measure the angle of the rock surface, in order to hunt for other surfaces of a similar age nearby. Matthews, dressed in a matronly white sun hat, used what looked like a jeweler's loupe to search for evidence of zircon crystals, which could be used to radiometrically date the rock. Few of these surfaces had ever been systematically dated, in part because the zircon extraction process is extremely tedious and costly. Matthews tried to explain the process to me in terms I could understand.

"First I take the rock and I break it into tiny little bite-sized pieces, then I mill them down to powder. Then I sieve the powder. Then I mix that powder with water and put that over what's called the Roger's Table, which works on the same principle as panning for gold. I just sit there for hours with a big bucketful, spooning one tablespoonful at a time. The table jiggles, and all the dense minerals go to the very end of the table and all the light clays go to the side. Then I do that all over again. That takes a day in itself. Then there's a technique called Frantzing, where you slowly crank up the strength of a magnet and slide the minerals down tiny little chutes. Different minerals are magnetically attracted at different strengths, so some of them get picked up. In the last stage, you use a horrible, nasty chemical called methylene iodide, which is a 'heavy liquid,' in that it's a lot denser than water but has the same viscosity, which means that things that would normally sink in water float in it. And because zircon is particularly dense, it sinks while everything else floats up. Then I pipe that up and squirt it onto a piece of filter paper. You dry this piece of paper out, after spending three days bashing this rock to buggery, and then you put it under the microscope and you *pray* that there's something under there."

He sighed like a man playing a game with terrible odds, but one he nevertheless enjoyed. "So I might start with a rock sample half as big as my backpack and end up with maybe forty zircon crystals so small that you can't see them." The crystals would then be worn down with a strong acid, then measured to determine how much of the zircon's uranium had decayed into lead, which would give an indication of its age, give or take a few hundred thousand years.

A few hours later we made our way over to the area's most famous fossil bed, the blandly named D Surface, which cantilevers out high over the ocean. Before we stepped out onto the bedding plane, we removed our shoes and put on polyester booties to protect the fossils from erosion. It felt like a ritual act, as if we were stepping into a temple.

The rock was huge and flat and intricately patterned, like the floor of a mosque. After visiting a lesser bedding plane, in which I often had to squint and tilt my head to make out what was fossil and what was figment, the profusion and sharpness of the fossils on D Surface was astounding. The Pigeon Cove surface had held about fifty fossils; this one held 1,500. They were everywhere, a vast fossilized garden of fronds and blobs and spirals, some larger than a large hand.

Of course, it was not an actual garden; plants would not appear in the fossil record for another two hundred million years. For some

reason I was stuck on this point. They *looked* like plants, I kept saying. Matthews explained that this was because, this far in the past,
the lines between the kingdoms grow fuzzy. We, and every organism
currently living on earth, he said, are at the crown of the tree of life.
Down at the base of the tree lie the very first single-celled organisms,
from which everything else sprang. So the further down the trunk
of the evolutionary tree you look, the more organisms resemble
one another. "That's when you get into the nitty-gritty definitions of
what defines, say, an animal and a fungus," he said. "They're actually
biologically really close, but they just 'decide' to stick their cells together slightly differently. And just because one evolved to stick its
cells together differently than another, one mainly just grows on dead
trees, and the other has conquered the earth."

What, then, makes a conqueror? We have sex. We eat life, not
sunlight. We contain multiple cells, which in turn contain nuclei, but
lack rigid walls. And, in almost every case, we grow muscles.

Muscles, I learned, are a crucial component of Liu's big question.
While many kinds of organisms (even single-celled ones) can swim,
reach, float, squirm, and even roll, only animals have developed muscle fiber, which has allowed us to move in a wider variety of ways and
heave around vastly more weight. Liu's trails, then, could help unravel
the question of when animal life began. Because if something was
big and strong enough to create those trails 565 million years ago, it
must have had muscles, which means it must have been an animal.

In a neat coincidence, the same summer Liu discovered the fossil
trails, he also unearthed a brand-new Ediacaran species with notice-
able muscle fibers—at 560 million years old, by far the earliest muscles in the fossil record. While he doesn't believe it was responsible
for making the trails, it does provide evidence that musculature was
developed earlier than anyone had previously thought. The new species was a ghastly-looking thing, a webbed, cupped hand reaching up
from a slender stalk, as if waiting to trap a passing foot. Liu named

it *Haootia quadriformis*, drawing from the language of the island's indigenous inhabitants, the Beothuk. *Haoot* means, simply, "demon."

+

Just as life on Earth requires both reproduction and death for it to evolve, the growth of science requires not just the birth of new discoveries, but the death of old ones. Any new scientific discovery is open to attack—the bigger the finding, typically, the fiercer the attack. Shortly after Liu published the paper outlining his discovery of the world's oldest fossil trails in 2010, Greg Retallack, a professor specializing in paleopedology (the study of fossil soils), attempted to debunk Liu's findings. Retallack claimed that the trails were not the result of animals, but rather "tilting traces," the marks of pebbles being washed about by the tide. Liu published a swift response addressing each of Retallack's points. Then he invited Andreas Wetzel, the German ichnologist who first introduced the notion of a "tilting trace," to view the fossil trails in person. Wetzel assured Liu they were not tilting traces.

Around the same time another paper emerged, from a University of Alberta team working in Uruguay, that claimed to have found trails that were twenty million years older than Liu's. This paper was challenged by a team of Uruguayan geologists who argued that the rocks had been dated incorrectly, and that similar fossils had only been found in much younger, Permian rocks. Casting further doubt on their discovery was the fact that the trails were significantly older than Liu's yet relied on a trail-maker with a vastly more advanced body structure. This discrepancy is akin to an automotive historian claiming to have uncovered a flying car from the nineteenth century. It's not impossible, just unlikely. (But then, Liu charitably pointed out to me, his discovery had also once seemed unlikely.)

Such is the gladiatorial, or perhaps more accurately, Darwinian, nature of research science. The goal, as famously explicated by the philosopher Karl Popper, is that in the competition for funding and fame, any false research will be falsified, and only the strongest theories will survive. However, an unfortunate side effect of this dynamic is what Martin Brasier called the MOFAOTYOF ("My Oldest Fossils Are Older Than Your Oldest Fossils") principle: "The tendency among all scientists, and certainly among all journalists, is to make their scientific claims as strong as they possibly can from the limited amount of material available." Bold conjectures are an integral part of healthy science, just as one initially underbids when negotiating at a flea market so as to eventually reach a fair price. But this tendency to exaggerate can prove dangerous, especially when the results trickle out to the general public, who, not understanding that falsification is a necessary part of the game, can develop a jaundiced view of any and all new scientific claims.

When I spoke to Brasier over the phone in 2013, he told me that the uncertainty inherent in this field of research was its appeal: He believed pure science is to be found on the edge of the darkness. "Karl Popper would have said that astrophysics and paleontology are not real science because you can't go out and sample it," he told me. "I think absolutely the opposite. I think this is actually where science is. It's trying to guess what lies over the hill and map terra incognita. When people come in and colonize, that's just technology." Brasier believed a scientist was, at heart, an explorer.

One of the strange side effects of working at the edge of the known universe, as Liu does, is that the more you learn, the more uncertain things become. As I talked with Liu and his team, I was constantly unlearning old assumptions I had held; even basic, bedrock knowledge began to disintegrate. What, for example, is the definition of movement? (Does floating count, or must one propel oneself? If so, with what kinds of tissues?) Is "animal" a clear-cut category, or a fuzzy-edged one? What, moreover, does it mean to be a living thing at all?

Life, according to Mikhail A. Fedonkin's *The Rise of the Animals*, a touchstone text among Ediacaran researchers, is defined merely as "a self-perpetuating chemical reaction" or "a self-assembling dynamic system." The fundamental element of this system is the membrane. Without membranes, there are no cells, and without cells there is no discrete space for chemical reactions to perpetuate. "The cell membrane also allows communication with the outside world, but regulates what comes in and what goes out," wrote Fedonkin. The communication is imperfect, but that imperfection is what defines one cell from another.

For billions of years, these single cells were the only living things on Earth. However, a cell benefits from better communication and cooperation with other cells. So some cells may have formed symbiotic relationships with others, then gathered into colonies, and, eventually, bound together into tissues. Interdependence both shackles and strengthens. Despite the restriction of freedom, tissues allow a much greater range of body types, including bilateral symmetry (a distinct front end and back end), which is the structural basis for the wild array of beasts that now stalks the earth and sea and sky. Matthews cheekily summed up this billion-year-long process of evolution from cells to bilaterians thus: "Tissues developed because it's nice to have muscles and an ass. It's not good to shit out of your face. It's just not a very good idea."

We tissuey beings define our individual selves as enclosed, tight-knit systems. ("Otherwise," Matthews said, "your arm would run away.") But here too our assumptions begin to break down. As we sat eating our picnic lunches on a flat rock overlooking the sea, Stewart mentioned that he had recently been reading an article in a science magazine that asked what truly defines a human being, since our bodies are dependent on an unseen universe of microorganisms to survive. There are, for example, at least as many bacterial cells as human cells in the human body—possibly many more.

"There are more cells *in* you that *aren't* you," Matthews said.

"Yeah," said Stewart. "Which sort of brings you to a point of 'What *is* you? What *am* I?' I had an existential crisis while I was reading it."

As I chewed my plum down to its wet-furred pit, a druggy feeling overtook me. I suddenly became aware of my own complexity: a riverine inner landscape swimming with cells both native and foreign; varied tissues clinging to and pulling against an architecture of bone; a digestive tract breaking down a bolus of plant material; two feet pressed against the earth; two nostrils sucking and spouting air; and in between, a branching network of nerves flickering with electricity in a furious effort to make sense of it all. Inside the human body lies a realm of perpetual darkness and riotous life, much of it still unexplored. We are, each of us, wild to our marrow.

+

After lunch, we headed east along the shoreline, ascending through geologic time over younger and younger rocks. We were following a stratigraphic chart of Liu's, which looked like a multi-tiered ice-cream sandwich. Each layer depicted a stratum of rock; embedded in some, like a sprinkling of chocolate chips, were known fossils. We paused on one surface where the chart indicated there should be an array of disc-shaped fossils, but we couldn't see them. The angle of the light wasn't right; certain layers reveal their fossils only when the sun is low and shadows become pronounced. Liu got down on his haunches, his eyes scanning the rock. Then they clicked into focus. "Ooh," he said, pointing. "There's one." I followed his finger. The lines of an ovoid fossil emerged from the pixelated background, like in those Magic Eye books I used to go cross-eyed over as a child.

"There are loads of them, actually!" Liu said, his hand sweeping across the marbled surface. As his finger passed over them, the outline of a half-dozen other discs rose from the rock.

I was baffled. When I looked at these rocks, I saw a jumbled code:

```
QAZXSWEDEWASDXDRTHUJKGFRTDXDRTDEASVBS
EDEGFRTGFDEWRDSAWEDSERTGVFTYHGYUDXKPT
DRTGFRTJHUJKMJUIOLKMJIOLKOPOIUJHHYYHGH
TFTFGFFRDCVFDFEDSXDEDFDCVGFASOEGFGHJF
GFGFDFDSXSWWSAZXDXCFDXCFRDSFOSSILFGYST
GLJXPYFOSFOSSILHYHUIOPLKJUYTGHNBVGFTRD
VCFKIUJASOPOIKMJNHINJUHYNJHNJMKIJUHNJM
NMJHJKJHBNJNMKJNMKJNMKJMKMJHBEAJDAEI
IEODFKDLSDKFJMCLXSOEOEOEKRJFIKDOLSXCKM
JDLSKOGHNHJUFOEJOABARIDONGOEOIDNOODC
OSOIDKEDINTOQKIOPREDEWASDXDRTGFRTGFDE
RDSAWEDSERTGVFTYHGYUJHUJKMJUIOLKMJIOLK
POUJHHYYHGHYGTFTFGFFRDCVFDFEDSXDEDFDC
VGFASOEGFGHHGFGFDFDSXSWWSAZXDXUIPCFDX
CFRDSFOGOSTFGYSTOGOOOFOHFOSSILYHUHUM
PLKJUYTGHRDFGBVCFKIUJASOPOIKMJNHHNJUHYL
HNJMKIJUHNJMKJHNMHJKJHBNJNMKJNMKJNMK
MKMJHBEAJDAEIURIEODFOSSILDKFMCLXSOEOEO
KRJFKDOLSXCKMKFJDLSKOGHNHJUFOEJOABARD
NGOEOIDNOODCNOSOIDKEIDINTOAQNBVCXEDEH
SDXDRTGFRTGFDEWRIDSAWEDSERTGVFTYHGYUJ
KMJUIOLKMJIOLKOPOIUJHHYYHGHYGTFTFGFFRD
DFEDSXDEDFDCVGFASOEGFGHGFGHGFGFDFDSXS
WSAZXDXCFDXCFRDSFOGOSGOOOFOSSILSFHYHU
PLKJUYTGHNBVGFTRDFGBVCFKIUJASOPOIKMJNH
UHYNJHNJMKIJUHNJMKJHNMJHJKJHBNJNMKJNM
NMKJMKMJHBEAJDAEIURIEODFKDLSDKFJMCLXSO
RDFGBVCFKIUJBARIDONGOEOEOEKRJFKDOLSXCK
KFJDFKDLSDKFOSSILSKOEOEOEKRJFKDOLSXCKM
JDLSKOGHNHJUFOEJOABARIDONGOEOIDNOODC
OSODKEIDINTOQKOPREDEWASDXDRTGFRTGFDEW
```

But when Liu looked, a clear picture arose:

```
XXXXXXXXXXXXXXXXXXXXXXXXXXXXXXXXXXXXXXXXXXX
XXXXXXXXXXXXXXXXXXXXXXXXXXXXXXXXXXXXXXXXXXX
XXXXXXXXXXXXXXXXXXXXXXXXXXXXXXXXXXXXXXXXXXX
XXXXXXXXXXXXXXXXXXXXXXXXXXXXXXXXXXXXXXXXXXX
XXXXXXXXXXXXXXXXXXXXXXXXXXXXXFOSSILXXXXX
XXXXXXXXXFOSSILXXXXXXXXXXXXXXXXXXXXXXXXX
XXXXXXXXXXXXXXXXXXXXXXXXXXXXXXXXXXXXXXXXXXX
XXXXXXXXXXXXXXXXXXXXXXXXXXXXXXXXXXXXXXXXXXX
XXXXXXXXXXXXXXXXXXXXXXXXXXXXXXXXXXXXXXXXXXX
XXXXXXXXXXXXXXXXXXXXXXXXXXXXXXXXXXXXXXXXXXX
XXXXXXXXXXXXXXXXXXXXXXXXXXXXXXXXXXXXXXXXXXX
XXXXXXXXXXXXXXXXXXXXXXXXXXXXXXXXXXXXXXXXXXX
XXXXXXXXXXXXXXXXXXXXXXXXXXXXXXXXXXXXXXXXXXX
XXXXXXXXXXXXXXXXXXXXXXXXXXXXXXXXXXXXXXXXXXX
XXXXXXXXXXXXXXXXXXXXXXXXXXXXXFOSSILXXXXX
XXXXXXXXXXXXXXXXXXXXXXXXXXXXXXXXXXXXXXXXXXX
XXXXXXXXXXXXXXXXXXXXXXXXXXXXXXXXXXXXXXXXXXX
XXXXXXXXXXXXXXXXXXXFOSSILXXXXXXXXXXXXXXX
XXXXXXXXXXXXXXXXXXXXXXXXXXXXXXXXXXXXXXXXXXX
XXXXXXXXXXXXXXXXXXXXXXXXXXXXXXXXXXXXXXXXXXX
XXXXXXXXXXXXXXXXXXXXXXXXXXXXXXXXXXXXXXXXXXX
XXXXXXXXXXXXXXXXXXXXXXXXXXXXXXXXXXXXXXXXXXX
XXXXXXXXXXXXXXXXXXXXXXXXXXXXXXXXXXXXXXXXXXX
XXXXXXXXXXXXXXXXXXXXXXXXXXXXXFOSSILXXXXX
XXXXXXXXXXXXXXXXXXXXXXXXXXXXXXXXXXXXXXXXXXX
XXXXXXXXXXXXXXXXXXXXXXXXXXXXXXXXXXXXXXXXXXX
XXXXXXXXXXXXXXXXXXXXXXXXXXXXXXXXXXXXXXXXXXX
XXXXXXXXXXXXXXXXXXXXXXXXXXXXXXXXXXXXXXXXXXX
XXXXXXXXXXXFOSSILXXXXXXXXXXXXXXXXXXXXXXX
XXXXXXXXXXXXXXXXXXXXXXXXXXXXXXXXXXXXXXXXXXX
XXXXXXXXXXXXXXXXXXXXXXXXXXXXXXXXXXXXXXXXXXX
```

When I asked him how exactly he managed to zero in on the relevant bits of visual information, Liu said that the secret was training your eye. Stewart disagreed. He thought that Liu was born with better eyes than most—not just the physical eye, but the whole perceptual apparatus. "This guy does it to me all the time," Stewart said. "We'll go fossil hunting in England, and he's like 'There's one. There's one.' I get so frustrated. But there's a reason he's doing this for a living."

Later, Liu obliged to describe the process in greater detail. The key, he said, lay in "trying to cut out all the noise"—recognizing and eliminating any nonbiological features that might resemble fossils to the untrained eye. Once de-cluttered, the pattern can emerge. You also need a "search image," he added. "If you know to look for something, you'll see it. But if you don't, you can miss it."

It struck me that a similar problem plagues all branches of science. Scientists looking for new discoveries appear to be trapped in a logical bind: it is exceedingly difficult to find something when you don't even know what it is you are looking for. Here is where the imaginative work of hypothesizing enters into play. By extrapolating from known patterns, we can make predictions and envision new phenomena, which we can then begin looking for in the world.

Hypothesizing is such a powerful tool that it can bear fruit even when the things we think we are looking for are not, in fact, there. Huxley, foreshadowing Kuhn's theory of paradigm shifts by some fifty years, argued that this speculative process guides all scientific inquiry:

> Man approaches the unattainable truth through a succession of errors. Confronted by the strange complexity of things, he invents, quite arbitrarily, a simple hypothesis to explain and justify the world. Having invented, he proceeds to act and think in terms of this hypothesis, as though it were correct. Experience gradually shows him where his hypothesis is unsatisfactory

and how it should be modified. Thus, great scientific discoveries have been made by men seeking to verify quite erroneous theories about the nature of things. The discoveries have necessitated a modification of the original hypotheses, and further discoveries have been made in the effort to verify the modifications—discoveries which, in their turn, have led to yet further modifications. And so on, indefinitely.

The open-ended nature of science is either its greatest asset or its fatal weakness, depending on one's outlook. Those of a mystical or skeptical cast of mind cite its mutability as proof that all scientific knowledge is ultimately shallow and illusory, while a believer in the scientific method finds comfort in the fact that it continually evolves to more tightly fit the contours of the universe's dark terrain.

+

Our rise through geologic time ended at the bedding plane that bore Liu's fossil trails. On a rock wall facing the sea there protruded a waist-high shelf. We hovered over the shelf, looking down. Once again, I saw only a flat expanse of stone until Liu pointed out the trails subtly etched into the rock.

Here, finally, was what I had come to see: the world's oldest trails. They were easy to miss; it looked as if someone had lightly dragged a pencil eraser through drying concrete. Matthews opened his canteen and poured some water over the rock, so the trails would stand out in starker relief. Even still, I came to understand how dozens of other paleontologists had failed to notice them. All around were large, distinct body fossils impressed into grand sweeping surfaces. Liu's trails were like a poem carved onto a handrail in a stairway of the Louvre.

We worked our way along the shelf, inspecting yet more trails. Some were larger than others, but none were wider than a thumbprint. Most were relatively straight, but one peculiar trail looped back on itself, like a snake in agony. Liu believed that it provided further evidence that the marks were not, as Retallack had argued, produced by a rock or shell being dragged by a current along the seafloor.

I lightly ran my fingers over the trails. They bore the distinct texture of life. Their surface was patterned with a series of nesting arcs: ((((((((. Liu thinks each arc was made by the creature's circular foot as it inflated with water and extended forward, smearing the front edge of the previous impression. At the end of some of the trails was a small dimple—(((((((()—called a "terminal impression," which might indicate the organism's final resting place.

Modern sea anemones creep along the seafloor using a similar system of hydrostatic inflation. And this, Liu thought, could provide a clue as to why the first animals made trails. Many of the Ediacarans found on Mistaken Point were believed to have lived their lives secured to the ground by suction cup–like feet, with their fleshy bodies extending out into the water column to gather food. Modern animals with similar body types typically prefer to latch on to a hard substrate, like stone or, when available, glass. In his lab, Liu had observed that when sea anemones were forcefully pried loose from the aquarium's glass, they would creep across the tank's sandy bottom until they encountered another hard, flat surface.

Liu's best guess was that his fossil trails were similarly formed: an Ediacaran was washed from its rock and, mired in loose sediment, it struggled through the muck to regain its perch.

I had come to Mistaken Point hoping to gain some understanding of why the first animals began to roam. I would have assumed the ur-trail-maker was propelled by either food, sex, or imminent danger. I hadn't accounted for this counterintuitive but perhaps equally primal need: the desire for stability.

I thought back to my experience of being lost amid the Tuckamore, how intensely I had yearned for the comfort of a building, or even just a trail—something solid and familiar to which I could cling. Huxley had felt that yearning too; I suspect most people have. There is no sure way of knowing what the ancient Ediacarans felt, or if they even *could* feel. But here, written in stone, was a clue. In the end—or rather, in the beginning—the first animals to summon the strength to venture forth may simply have wanted to go back home.

CHAPTER 2

RETURNED HOME from Newfoundland with a skull full of new questions. For reasons I couldn't quite grasp, the fossil trails at Mistaken Point continued to vex me. The more I thought about those inscrutable old scribbles, the more they struck me as curiously inert—and not only because their makers had been dead for half a billion years. Trails tend to possess a certain vital suppleness, a formal litheness or grace, which they altogether lacked.

It was only later, by studying the invisible trails of ants—arguably the world's greatest trail-makers—that I finally located the glitch: strictly speaking, the Ediacaran trails were not really trails, they were *traces*. Ant trails gain their magical efficiency from a very simple feedback mechanism: a trail is left behind by one ant and then followed by another, and another, and another, subtly evolving with each subsequent trip. We have no reason to believe that one Ediacaran would have followed in the footsteps (or rather, foot smears) of another. Their trails were a call without a response.

The words we English speakers use to describe lines of

movement—trails, traces, tracks, ways, roads, paths—have grown entangled over the years. I am as guilty of this conflation as anyone else, in part because the meanings of these words, much like the things they denote, tend to overlap. But to better understand how trails function, it helps to momentarily tease them apart. The connotations of *trail* and *path*, for example, differ slightly: a "path" sounds dignified, august, and a bit tame, while a "trail" seems unplanned, unkempt, unruly. The *Oxford English Dictionary* editors define a trail, rather sniffily, as "a rude path." As they point out, trails only ever pass through wild regions, never cultivated ones; it would sound awkward to speak of strolling down a "garden trail." But why?

When we take a step back, we find that the key difference between a trail and a path is directional: paths extend forward, whereas trails extend backward. (The importance of this distinction becomes paramount when you consider the prospect of lying down in the *path* of a charging elephant versus lying down in its *trail*.) Paths are perceived as being more civilized in part because of their resemblance to other urban architectural projects: They are lines projected forward in space by the intellect and constructed with those noble appendages, the hands. By contrast, trails tend to form in reverse, messily, from the passage of dirty feet.

Over time, the meaning of the two terms converged in North America in the nineteenth century, when Anglos often found themselves traveling almost exclusively on trails left behind by animals and Native Americans. The word acquired its flavor out west; the *OED*'s earliest citation of "trail"—meaning a footpath, animal trace, or wagon road—dates back to the Lewis and Clark expedition. Colonel Richard Irving Dodge, in 1876's *Plains of the Great West*, drew from his tracking experience to give us this helpful definition: trails are a string of "sign" that can be reliably followed. I like this definition, because it gets us away from the erroneous assumption that a trail is synonymous with a strip of bare dirt, but it requires some explanation. "Sign"—a word, like its synonym "spoor," that is always written in the singular—refers to the marks left behind by an animal in its passing: footprints, droppings, broken branches, tree trunks rubbed bare by antlers. "A trail is made up of 'sign;' but 'sign' is, by no means, a trail," Dodge clarified. "Deer make 'sign,' but it may be impossible to trail them." Trails, in this—albeit, somewhat tautological—formulation, are simply *that which can be trailed.*

Something miraculous happens when a trail is trailed. The inert line is transformed into a legible sign system, which allows animals to lead one another, as if telepathically, across long distances. (These signs can be physical, chemical, electronic, or theoretical. The medium, in this case, is not the message.)

The truly incredible thing about these sign systems is that they require no special intelligence to create or follow. Some of the animal kingdom's earliest trail followers were likely marine gastropods (the ancient progenitors of snails and slugs), which emerged during the Ordovician period. Modern-day marine gastropods regularly track one another's slime trails across the seafloor by tasting the trail's mucousy surface, a process called "contact chemoreception." The slime of gastropods primarily speeds their travel, but this slick medium has also evolved into a signaling mechanism, much like how the smooth surface of the highway, as opposed to the bumpiness of the shoulder, signals that a driver hasn't veered off the road. Certain gastropods, like mud snails, only travel forward on slime trails—following a gradient toward the freshest mucus—which allows them to shadow one another on their herd-like migrations. Limpets, on the other hand, secrete trails that they trace in reverse, groping their way back to the nook-like homes they carve into rocks.

Slime trails function equally well on land, where terrestrial snails and slugs make frequent use of them. Darwin relayed the story of an acquaintance, named Lonsdale, who once placed two Burgundy snails in a "small and ill-provided" garden. The stronger of the two snails climbed over a wall into an adjoining garden where there was more to eat. "Mr. Lonsdale concluded that it had deserted its sickly mate," Darwin wrote, "but after an absence of twenty-four hours it returned, and apparently communicated the result of its successful exploration, for both then started along the same track and disappeared over the wall."

Mr. Lonsdale seems not to have considered the possibility that, if the first snail can leave behind an intelligible trail for its mate to follow, no other form of communication is necessary. The trail provides one of the animal kingdom's most elegant ways to share information. Each inch is a sign, like a scrawled arrow, reading simply:

This way . . .

This way . . .

This way . . .

+

The invention of trails provided a powerful new tool of animal communication, a kind of proto-Internet capable of running on a simple binary language—this way and not. No species has exploited this new technology more brilliantly than ants, which routinely use trails to find new sources of food and transport it back to the nest. Scientists now study these tiny, but stunningly efficient, trail systems to learn how to more quickly route bits of information through our own fiber-optic networks.

For many centuries, it was a mystery how ants were able to organize themselves so deftly. Some believed each ant was possessed of a tiny special intelligence, which afforded it rationality, language, and the ability to learn, as the naturalist Jean Pierre Huber argued in 1810. Put simply, this view held that ants found their way to food using their wits, and then "spread the word" of their findings throughout the nest. (This highly anthropomorphized notion remains prevalent among folktales and children's stories, from *Aesop's Fables* to T. H. White's *The Once and Future King.* In many of these renditions, like White's, the worker ants are given their marching orders by an all-powerful "queen.") Opponents of this theory, following the teachings of Descartes, held that ants were possessed of no intelligence whatsoever—or, in the language of the time, no "soul"—but were mere machines directed by an almighty deity who either manipulated them like marionettes or engineered them like windup toys. The naturalist Jean-Henri Fabre, an unfashionably late proponent of this theory, wrote in 1879, "Can the insect have acquired its skill gradually, from generation to generation, by a long series of casual experiments, of blind gropings? Can such order be born of chaos; such foresight of hazard; such wisdom of stupidity?" Fabre concluded that it could not. "The more I see and the more I observe, the more does this [divine] Intelligence shine behind the mystery of things."

On the one side of this debate lie insects blessed with individual wit, on the other, insects cursed with perfect idiocy, but steered by an omniscient hand. It was not until very recently that scientists began to understand that the answer lay somewhere in between: the complex behavior of ants arises not from smart individuals, but from smart systems—a form of wisdom that exists *between*, as well as inside, living things.

All animals fall somewhere on the spectrum between internalized and externalized intelligence. At one extreme of this spectrum lies the mountain hermit, thoughts swirling about in his lonely head like moths in a bell jar. At the other end lies the slime mold. As sprawling, single-celled blobs, slime molds are about as stupid as an organism can be: they lack even the most basic rudiments of a nervous system. However, they have nevertheless developed a very effective technique to hunt for food: They extend their tentacle-like pseudopods, grope around, and then retract them whenever they come up empty. As they retract, the pseudopods leave behind a trail—or rather, a kind of *anti-trail*—of slime indicating where food has not been found. Then, continuing their blind search, they head off in a new, slime-free direction. Using roughly this same trial-and-error method, slime molds can solve surprisingly complex problems. When researchers tasked a slime mold with connecting a series of oat clusters mirroring the location of the major population centers surrounding Tokyo, the slime mold effectively re-created the layout of the city's railway system. Linger a moment over that fact: A single-celled organism can design a railway system just as adroitly as Japan's top engineers. Whatever intelligence slime molds have, though, is wholly external. When their enclosure is wiped down evenly with slime—effectively erasing their trails—slime molds will begin to wander aimlessly, as if struck with dementia. They don't retain any information; the trail does.

As a species, humans straddle a line between external and internal intelligence. With big brains and (typically) small clan size, humans have traditionally harnessed individual cleverness to outcompete rivals for food and mates, to hunt and dominate other species, and, eventually, to seize control of the planet. As later chapters will show, we have also externalized our wisdom in the form of trails, oral storytelling, written texts, art, maps, and much more recently, electronic data. Nevertheless, even in the Internet era, we still romanticize the lone genius. Most of us—especially us Americans—like to consider any brilliance we may possess, and the accomplishments that have sprung from it, as being solely our own. In our egotism, we have long remained blind to the communal infrastructure that undergirds our own eureka moments. This egotism extends to our regard for pathways: when we write about trails, we tend to describe them as the creation of a single "trailblazer," whether it is Daniel Boone blazing the Wilderness Road or Benton MacKaye dreaming up the Appalachian Trail. The reality of how most trails form—collectively, organically, without the need of a designer or a despot—has been increasingly apparent to scientists for centuries, but has remained invisible to most of us for far too long.

+

The story of how we grew wise to the wisdom of insect trails begins, oddly enough, with the lowly caterpillar. One spring day in 1738 a young Genevan philosophy student named Charles Bonnet, while walking through the countryside near his family's home in Thônex, found a small, white, silken nest strung up in the branches of a hawthorn tree. Inside the nest were squirms of newly hatched tent caterpillars, which bristled with fiery red hairs.

At just eighteen years old, frail, asthmatic, myopic, and hard of hearing, Bonnet was a somewhat unlikely naturalist. But he was blessed with patience, attentiveness, and a relentless, burning curiosity. As he approached the cusp of adulthood, his father had begun pressuring him to become a lawyer, but he wanted to spend his life exploring the microcosmos of insects and other tiny creatures, a profession that had scarcely yet been invented.

Bonnet decided to cut down the hawthorn branch and carry it back home. At the time, most naturalists would have sealed the caterpillars in a powder jar, called a *poudrier,* to better inspect their anatomy. But Bonnet wanted to observe the caterpillars' natural behavior wholly unobstructed, *en plein air,* yet from the comfort of his home. He struck upon the idea of mounting the hawthorn branch outside the window frame of his study. That window soon became a kind of antique television, a glass screen displaying a miniaturized world, before which he spent countless rapt hours.

After two days of patiently waiting for signs of life, Bonnet watched the caterpillars emerge from their nest and begin to march in single file up the windowpane. After four hours, the procession had successfully scaled the window; then it turned around. In descending, strangely, the caterpillars followed the exact path they had climbed. Bonnet later wrote that he even traced their route— presumably, with a wax pencil on the windowpane—to see if they ever deviated from it. "But they always followed it, faithfully," he wrote.

Each day Bonnet watched as the caterpillars mounted exploratory expeditions across the windowpane. Paying closer attention, he noticed that as they crawled, each caterpillar laid down an ultra-fine white thread, which the others followed. Curious, Bonnet rubbed his finger across their trail, breaking the thread. When the leader of the returning party arrived at the rupture, it turned back, apparently confused. The one behind it did the same, and the one behind that. Each subsequent caterpillar plodded calmly along until it reached the gap in the trail, at which point it either turned around or stopped to feel about for the thread, like a man groping for a dropped flashlight. Finally, one of the caterpillars, which Bonnet deemed "hardier than the others," dared to venture forward: a thread was extended across the void, and the others followed.

Emboldened, Bonnet collected more caterpillar nests, which he placed on his mantel. Soon, scores of caterpillars were exploring his bedroom, meandering across the walls, the floor, even the furniture. Feeling, no doubt, like a small new god, Bonnet found he could control where the caterpillars traveled simply by erasing certain trails. He delighted in showing this trick to visitors. "You see these little caterpillars who walk in such good order?" he would ask. "Well, I bet you that they will not pass beyond this mark"—and he would swipe his finger across their route, stopping them cold.

+

Along the southern stretches of the Appalachian Trail, I too sometimes encountered mysterious little white tents in the crotches of trees. Occasionally they grew to monstrous proportions; I would turn a corner to find a tree wholly enveloped in a polygonal cloud. "Mummy trees," my fellow hikers called them.

For a reason I couldn't quite place, they gave me the shivers. Tent caterpillars, I would later learn, are essentially creepy animals. Their faces resemble black masks, and their bodies are quilled over in fine, toxin-tipped spines, which can detach and float for more than a mile on a windy day, causing rashes, coughing fits, and pink eye. Some species of tent caterpillar undergo rampant population booms on a ten-year cycle, covering the countryside like spilled oil. In June 1913 a stream of forest tent caterpillars climbed up onto the tracks of the Long Island Rail Road; the rails were soon so thickly slathered with their remains that the wheels of approaching trains spun in place.

A biologist named Emma Despland once told me about the time she walked into a stand of sugar maples during a tent caterpillar outbreak. She described it as a "ghost forest."

"It's June and there are no leaves on the trees, and there are these big strands of gunky silk, like Halloween decorations," she said. "And then you hear this rain falling. Except it's not rain. It's caterpillar poop."

Even among biologists, tent caterpillars are little loved. And yet for centuries, researchers like Despland have been studying them for one reason in particular: as consummate followers—perfectly faithful, perfectly foolish—tent caterpillars represent a reductio ad absurdum of what it means to follow a path. Despland told me that if you were to remove a younger caterpillar from its nest mates, it would spend all its time waving its head around in confusion, looking for a trail, and probably starve to death. Alone, they are utterly hopeless, and yet collectively they can denude entire forests.

Curious to see firsthand how such timorous creatures manage to bind together and thrive in the world, I took the bus to Montreal to visit Despland's lab, where she studies the forest tent caterpillar. When I arrived she peeled open a Tupperware container to show me the caterpillars: a smattering of fuzzy black critters, like mouse turds come to life. Then, on an old desktop computer in her office, Despland showed me time-lapse video of an experiment she had been conducting to determine how they find food. In the experiment, the caterpillars were placed in the middle of a cardboard runway. On the extreme left end of the cardboard strip was a Quaking Aspen leaf, which caterpillars especially love to eat. On the right was the leaf of a hybrid poplar, *Populus trichocarpa* × *P. deltoides* (clone H11-11), which they find unappetizing. The experiment was simple: It was as if a group of blindfolded children were placed in the middle of a long hallway, which held a piece of chocolate cake at one end and a pile of raw celery at the other. Asked to find and share the more delicious item, children could quickly solve the problem by splitting up and calling out to one another. But how would the caterpillars?

Displayed on Despland's computer monitor were five strips of cardboard, on which five experiments were being conducted simultaneously, but she directed my attention to the second from the bottom, where, over the course of many minutes, a group of caterpillars had mistakenly ventured over to the hybrid poplar leaf. Others followed their trails, and soon the whole nest was crawling on the broad green leaf, though they ate virtually none of it. For an uncomfortably long period of time they failed to correct this initial mistake. They followed their trail back to their "bivouac" (a silken pad, which they construct as a resting place) in the middle of the strip, and then back to the hybrid leaf on the right, but none ventured to the left, where the tasty aspen leaf lay. It seemed each caterpillar would continue to follow the others to the hybrid leaf, leaving more trails, and more feedback, forever.

I recalled a peculiar incident Bonnet had once described witnessing, in which a group of pine processionary caterpillars mistakenly formed a circular trail leading all the way around the rim of a ceramic vase. The details are scant, but it seems they continued marching around and around for at least a whole day. This same phenomenon was later famously observed by Jean-Henri Fabre: to his amazement, the caterpillars walked in circles for more than a week before they finally broke the ouroborosian loop and escaped. In *Pilgrim at Tinker Creek,* Annie Dillard recounts the horror she felt while reading

Fabre's portrait of these soulless, circling automatons. "It is the fixed that horrifies us," she wrote. "It is motion without direction, force without power, the aimless procession of caterpillars round the rim of a vase, and I hate it because at any moment I myself might step to that charmed and glistening thread."

Despland's caterpillars seemed to be caught in a similarly brainless loop. For more than an hour, the pattern continued: the caterpillars returned to their bivouac, returned to the hybrid leaf, and returned to the bivouac again. I began to squirm.

Eventually, a small contingent broke away and ventured off in the opposite direction. They traveled slowly, with excruciating hesitancy, inching, ducking, cowering, stalling, nudging one another forward, and frequently turning back. Despland guessed that their hesitancy springs from a genetic aversion to ending up away from the pack, alone, where they could get picked off by a bird.

By the end of the second hour, the scouting party had finally made it to the aspen leaf, and others subsequently followed the trail they had blazed. Despite their initial misstep, by hour four all the caterpillars had found the correct leaf and gnawed it to a husk.

The foraging technique of these caterpillars is remarkably simple, even idiotic, but it works. The fail-safe, Despland explained, is that hunger induces restlessness, which eventually compels them to abandon the well-worn trails and go looking for something else. "The leaders tend to be the hungry ones," she explained. "Because they're the ones who are willing to pay the cost."

+

One year after his initial caterpillar experiments, Charles Bonnet was outside hunting for a new batch of caterpillars when he happened across a prickly flower called a teasel, whose head harbored a colony of tiny red ants. Ever curious, he plucked the flower, carried it back to his study, and planted it upright in an open powder jar.

One day Bonnet returned to discover that a number of the ants had deserted the nest. Searching about, he found them marching up his wall to nibble the wood at the top of his window frame. In his journal, Bonnet described watching one ant as it climbed down the wall, up the side of the powder jar, and back to the nest. At the same time, two ants emerged from the teasel head and climbed to the top of the window frame, following precisely the same route that the other had just descended.

"Instantly, it came to my mind that these ants which I had in front of me, like the caterpillars, left a trace that directed them in their course," he recalled.

Of course, he knew that ants did not emit a thread. But they did give off a strong smell, which is sometimes described as being reminiscent of urine. (This odor lent ants their archaic name, "pismires," and later, "piss-ants.") The substance, Bonnet theorized, could "more or less adhere to objects they touch, and then act on their sense of smell." He compared those "invisible traces" with the trails of wildcats, which are imperceptible to humans but plain as blood to dogs.

His suspicion was easily tested: as before, he rubbed his finger across the ants' pathway. "Doing so, I broke the path on a width equal to that of my finger, and I saw precisely the same spectacle the caterpillars had given me: the ants were diverted, their walk was interrupted, and their confusion amused me for me some time."

Bonnet had stumbled on an elegant explanation for how ant trails form, which required neither powerful memories, strong eyesight, nor simple language (as Huber and Fabre later proposed). Bonnet theorized, correctly, that ants ordinarily follow trails that lead to their homes and to food sources. However, some ants wander off track, "attracted by certain smells or other sensations to us unknown," spawning new side roads. If that rogue ant finds food, it will leave a new trail on its return to the nest, and other ants will follow. So, wrote Bonnet, "a single ant can lead a large number of its companions to a certain place without any need of a particular language whereby it announces the discovery that it has just made."

Judging from his journals, Bonnet seems not to have realized how historic this discovery was. Scientists had long suspected that ants deposit chemicals when they walk; in the sixteenth century, two German botanists, Otto Brunfels and Hieronymus Bock, discovered that ants produce formic acid after noticing that a blue chicory flower, when thrown onto an anthill, turns a vivid red. But no one properly connected the dots until Bonnet.

Around the time of Bonnet's death in 1793, the zoologist Pierre André Latreille confirmed Bonnet's suspicion that ants sniff their way through the world. He learned this by amputating the antennae from a number of ants; at once, he wrote, they began wandering aimlessly about, as if in "a state of intoxication or a kind of madness." Then, in 1891, Sir John Lubbock, the English polymath, performed a ground-breaking series of experiments involving Y-shaped mazes, bridges, and rotating platforms. Through painstaking experimentation, he showed that *Lasius niger* ants navigate primarily by using scent trails.

In the late 1950s, E. O. Wilson solved the riddle by locating the gland in fire ants that secretes trail pheromones. He had a hunch that the trail substance resided somewhere in an ant's abdomen, so he split the abdomen open and, using a pair of sharpened watchmaker's forceps, carefully removed all the organs. Then he smeared each organ across a piece of glass. After each stroke, he checked to see if it had any effect on a nearby colony of ants. Line after line, organ after organ—the poison gland, the hindgut, the little blob of lipids called a "fat body"—prompted no response. Finally he smeared out a tiny, finger-shaped organ called Dufour's gland. "The response of the ants was explosive," Wilson later recalled. "As they ran along they swept their antennae from side to side, sampling the molecules evaporating and diffusing through the air. At the end of the trail they milled about in confusion, searching for the reward not there."

By the year 1960, our fuzzy understanding of ant trails had snapped into sharp focus. Two crucial new terms were born concurrently: a pair of German biologists coined the term *pheromone*—chemical triggers, or signals—and Pierre-Paul Grassé introduced the notion of "stigmergy." Stigmergy is a form of indirect communication and leaderless cooperation, using signals deposited in the environment. Termites, for example, organize their massive construction efforts stigmergically: there is no foreman, and no direct communication between the termites. Rather, the termites respond to a series of simple cues in the environment (*if dirt here, move dirt there*), which in turn impel them to further alter the environment. This behavioral feedback loop can result in structures of stunning efficiency and resilience, like the towering termite mounds of Australia, which, proportional to their makers, are three times taller than our highest skyscrapers. With a combination of pheromones and stigmergy, even the simplest insects could build labyrinthine trail systems.

In the 1970s a biologist named Terrence D. Fitzgerald, being familiar with Wilson's work, intuited that tent caterpillars might also use trail pheromones. At the time, biologists believed that tent caterpillars followed their nest mates' silk, which is expelled from their mouthparts, but he had a hunch that they were secreting trail pheromones onto the silk from their back ends, as ants do. So he folded a plain piece of paper in half and ran its edge along the underside of a caterpillar's abdomen. Then he unfolded the paper and placed some caterpillars on it. Sure enough, the caterpillars marched back and forth along that crease, following the invisible line of pheromones just as Wilson's fire ants had. (Like Wilson, Fitzgerald was later able to isolate and synthesize these trail pheromones.) This discovery lent a neat symmetry to the path of inquiry Bonnet had started: We learned ants follow pheromone trails by studying tent caterpillars, then we learned tent caterpillars deposit pheromone trails by dissecting ants.

It may seem odd, then, that neither Wilson nor Fitzgerald cites

Bonnet's discovery. In fact, many of Bonnet's writings, including the story of how he discovered the true nature of ant trails, have never been published in English. Though his career showed a promising start, it ultimately veered off on an ill-fated path. In his twenties, Bonnet became a celebrated naturalist: the first person to witness a virgin birth among plant lice, the first to describe regeneration among worms, the first to learn that caterpillars breathe through holes in their skin, and the first to prove that leaves exhale. Then, in a cruel twist, his vision began to cloud with cataracts. Unable to practice observational science, he turned to more cerebral fields, like philosophy, psychology, metaphysics, and theology. Much of the latter half of his life was spent trying to reconcile the confusing new findings of the biological sciences with his deep religious faith, which held that the world was divinely engineered. Bonnet's magnum opus—an all-encompassing theory of the universe called the "Great Chain of Being," which posited that all species were slowly progressing toward a state of perfection over the course of eons—had some influence on later evolutionary theorists like Jean-Baptiste Lamarck and Georges Cuvier. But in the broader span of scientific progress, it proved little more than a theoretical side road, which was later made obsolete by Darwin's theory of evolution by natural selection. By the end of his life, Bonnet's blindness caused him to suffer from phantasmagoric visual hallucinations, which are now known as Charles Bonnet syndrome.* Today, that syndrome is primarily what he is remembered for, when he is remembered at all.

+

---

* Oddly enough, Charles Bonnet syndrome is so named not because Bonnet suffered from it, but because he was the first person to *describe* it: his grandfather, Charles Lullin, had suffered from it decades earlier, and Bonnet wrote a case study about him. The fact that Bonnet later acquired the syndrome as well was merely an unhappy coincidence.

Every trail tells a story, but some trails tell it more eloquently than others. The trails of Despland's forest tent caterpillars, for example, are blunt—they are essentially able to shout just one phrase: *This way!* The trails of certain ant species are more sophisticated: they can whisper as well as shout. The strength of the chemical trail tells the colony how desirable the trail's destination is, which allows for more nuanced communication and nimbler collective decision making. Scientists have long pondered how ants, which are individually quite stupid, can behave so intelligently as a colony. "The reason is," E. O. Wilson once wrote, "that much of the 'spirit of the hive' is actually invisible—a complex of chemical signals we have only now begun to reveal."

Consider the fire ant: Once a scout has found a food source, excited by its discovery, on its return trip it presses its stinger to the ground to release a stream of pheromone, like ink from a fountain pen. The more food it finds, the more pheromone it deposits.[*] Other ants follow this trail to the food, and then they lay more trails home. So if there is a large store of food, the trail will emerge quickly and blaze bright (chemically speaking), which will attract more ants. Then as long as food remains, the trail will continue to draw more ants. But once the food runs out, the trail evaporates, and the ants gradually abandon it for another, stronger trail. This process neatly illustrates how stigmergy allows simple beings to arrive at elegant solutions to complex problems all on their own.

The basic mechanism at work here is the feedback loop: cause leads to effect (an ant finds food, and deposits a trail as it returns to the nest), then that effect becomes a new cause (that trail attracts more ants), which then leads to an amplified effect (they lay down their own trails, recruiting more ants), ad infinitum. Feedback loops can be divided into two types: the desirable kind, known as a *virtuous*

---

[*] Much of the apparent stupidity of forest tent caterpillars stems from their lack of this seemingly simple, but quite ingenious, innovation.

circle, such as when ants leave stronger and stronger trails to a food source; or the undesirable kind, called a *vicious circle*, like when a microphone is placed too close to an electronic amplifier, which allows minor sounds to self-amplify into those terrible, high-pitched shrieks familiar to any concertgoer. (Scientists used to poetically refer to the latter phenomenon as a "singing condition"; today, we simply call it feedback.)

In the circling of tent caterpillars Bonnet and Fabre both witnessed, in a strikingly literal form, how the same mechanism that gives rise to a virtuous circle can also give rise to a vicious one. The animal psychologist T. C. Schneirla witnessed this grim transformation in 1936, while working at a laboratory on an island in the middle of the Panama Canal. One morning, the resident cook, Rosa, approached Schneirla in a state of feverish excitement. She led him outside, where he found, on the cement walkway in front of the library, hundreds of army ants marching in a circle about four inches across.

Army ants, which are blind, rely heavily on pheromone trails to navigate the world. Most of the time, they march in thick raiding columns, consuming everything in their path, a habit that has garnered them the nickname "the Huns and Tartars of the insect world." Schneirla could tell that something had clearly gone wrong with this colony. Instead of a marching column, the swarming mass resembled a ragged vinyl record, with concentric black rings spinning frantically around a hollow center. The circle widened as the day wore on. In the afternoon, rain began to drum the pavement, which divided the mass of ants into two smaller vortices, each rotating until nightfall. The next morning, Schneirla awoke to find that most of the ants had died; those that remained continued to plod in slow, tragic circles. A few hours later, all were dead, and other species of scavenging ants had arrived to carry them away.

Schneirla was careful to point out that the doomed loop had most likely formed because the ants were walking on perfectly flat cement; otherwise, the undulations of the jungle floor might have disrupted it.

However, looping trails had been recorded under different conditions by other prominent scientists, like the entomologist William Morton Wheeler, who once watched a group of ants circle the base of a glass jar for forty-six hours. ("I have never seen a more astonishing exhibition of the limitations of instinct," he wrote.)

In 1921 the explorer-naturalist William Beebe described running across a colony of army ants marching in an enormous circle through the Guyanese jungle. Beebe followed the procession for a quarter mile, under buildings and over logs, only to find that their trail ended where it began. Astounded, he traced the crooked circle again and again. The procession continued to circumambulate for at least a full day, "tired, hopeless, bewildered, idiotic and thoughtless to the last." By the time that a few stragglers at last broke from it and wandered away, most had fallen dead from starvation, dehydration, or exhaustion.

"This peculiar calamity may be described as tragic in the classic meaning of the Greek drama," wrote Schneirla. "It arises, like Nemesis, out of the very aspects of the ant's nature which most plainly characterize its otherwise successful behavior."

Beebe was more succinct. "The masters of the jungle," he wrote, "had become their own mental prey."

+

There is a simple reason why we find the image of circling ants or caterpillars so troubling. The first instinct of humans who are lost in the wilderness is to cling to any trail they find and never leave it. Indeed, authorities on wilderness survival commonly recommend this tactic: "When you find a trail stay on it," declares a backpacking guide published by the U.S. Forest Service, in a section titled "If You Get Lost." A trail, the naturalist Ernest Ingersoll once wrote, is a "happy promise to the anxious heart that you are going *somewhere*, and are not aimlessly wandering in a circle." A circular trail, then, is a cruel trick, a breach of logic, almost a kind of black magic.

A few years ago, my partner and I moved from a small apartment in New York City to a small cabin in British Columbia. Behind our property stands a tall cedar forest, and behind that lie the cold green waters of the Georgia Strait. The cabin often startles visitors when they first see it. Our next-door neighbor, Johnny, a classical guitarist, built it in a fit of modernist whimsy; it looks like two railroad cars stacked one atop the other. The ground floor is made of polished concrete, and the windows are almost the size of the walls. The insulation is scant, the electricity is always cutting out, the garden is plagued with deer, and the nearest supermarket is a twenty-five-minute drive away, but it's quiet and the air is clean and there are plenty of walking trails nearby.

At the end of our dirt road, where it joins a bend in the main thoroughfare like a needle in the crook of an arm, there is a little trail leading off into the woods. Johnny informed us that it led to a place the locals call the Grassy Knoll: a soft green tuft atop a rock outcropping over the strait. It's a lovely perch, they say, to watch the sun set over the mountains of Vancouver Island. However, Johnny strongly advised us against staying that long, for fear we might get lost. "Even I get turned around in there," he said, "and I've lived here for twenty years." Another neighbor, Corey, told us that he'd once gotten lost while walking in the forest with his infant daughter. When the sun began to set, he felt the first electric touch of panic, an early sign of what psychologists call "woods shock," or what used to be called simply "bewilderment." He kept his wits and got out, but as he recounted the story one night around a campfire, I could see the feeling seep up, blackly, behind his eyes.

Remi and I were not worried. It was just a little provincial park, after all, only five hundred acres. If lost, one need only walk three miles in any direction to hit either the coast or a road. Setting off at about three o'clock, we walked down to the end of our street and ducked through the dark curtain of branches.

On the other side, the light clouded to the opacity of sea glass. We looked around, blinking, at a temple of riotous decay, evergreen, shade-blue. On the coasts of British Columbia, the prodigious rainfall, sunny summers, and rich soil thrust the trees upward; the taller ones shed their lower branches like the vestigia of a rocket ship. But eventually that which nourishes, topples. The trees fall to the ground quietly, with a huff, and there turn to moist brown crumbs. Everything, everywhere, is furred with moss and bearded with lichen. Slip on a wet root and you will fall, weirdly slowly, through the gray-green air, and the ground will rise up to receive you in its soft heft.

The trail wasn't built by the park service—some local do-gooders had apparently cleared it—which meant that it was less legible than it might otherwise be. The only trail markings were the occasional ribbon tied to a branch where the trail skirted a swamp. The paths tended to split and splice. Johnny had given us directions for finding the knoll: turn right at the first T-shaped fork in the path and keep left until you reach the shore. It seemed simple enough.

When we reached the first fork in the trail, Remi propped a stick up against a tree so we would have a point of reference in case we got lost. We turned right and followed the trail around in a wide arc, chatting happily, until we found ourselves standing at a fork in the path. There, off to the side, was the stick Remi had propped up against the tree. We had gone in a circle. Befuddled, we turned around and set off in the opposite direction this time, and, minutes later, found ourselves back at the stick again.

In *Roughing It*, Mark Twain recalls heading out into a snowstorm, bound for Carson City. A man named Ollendorff, bragging that his instinct was as sure as any compass, promised to lead the group. After half an hour of plodding through the snow on horseback, the men came upon fresh hoof-tracks in the snow. "I knew I was as dead certain as a compass, boys!" shouted Ollendorff. "Here we are, right in somebody's tracks that will hunt the way for us without any trouble."

The men began to trot along the tracks. Before long it became evident that they were gaining on whoever was up ahead, because the tracks grew more distinct.

We hurried along, and at the end of an hour the tracks looked still newer and fresher—but what surprised us was that the *number* of travelers in advance of us seemed to steadily increase. We wondered how so large a party came to be traveling at such a time and in such a solitude. Somebody suggested that it must be a company of soldiers from the fort, and so we accepted that solution and jogged along a little faster still, for they could not be far off now. But the tracks still multiplied, and we began to think the platoon of soldiers was miraculously expanding into a regiment—Ballou said they had already increased to five hundred! Presently he stopped his horse and said:

"Boys, these are our own tracks, and we've actually been circussing round and round in a circle for more than two hours out here in this blind desert!"

It has been thought for centuries that human beings have a natural tendency to walk in circles. In 1928, a biologist named Asa Schaeffer claimed to have shown experimentally that blindfolded people walk, run, swim, row, and drive automobiles in spiraling patterns, a phenomenon he attributed to a "spiral mechanism" in the brain. The navigator Harold Gatty believed that people circled because of simple biological asymmetry; one leg tends to be longer or stronger than the other. ("With regard to our anatomy," he wrote, "we are all of us unbalanced.") In 1896, the Norwegian biologist F. O. Guldberg argued that circling was one of the "general laws" of biology. He recounted stories of birds wheeling in front of light-houses, schools of fish whirling in the lamps of deep-sea divers, hares and foxes circling to escape hunters, and men lost in fog wandering in loops.

Guldberg didn't see circling as a form of error. The law of circular movement, he argued, assures that lost animals will always be able to find their way back to "the native place to which animals in the struggle for existence must so often return, be it the udder of the cow, the warmth-giving wings and the guiding experience of the hen, or the sheltering tree or bush chosen by maternal instinct." Whether we like it or not, he argued, we circle to find our way back to familiar ground.

In 2009, a researcher named Jan Souman decided to test the circling instinct. He equipped volunteers with GPS tracking devices and instructed them to walk in a straight line across unfamiliar terrain, both in the forests of Germany and the deserts of Tunisia. Without the aid of directional cues, like the sun, the subjects did tend to circle back on their own trails; that much is true. "It seems easy to walk in a straight line," Souman told me. "But if you think about it, it's actually not that easy at all." Like riding a bicycle, walking a straight line is in fact a complex neural balancing act, which is what makes it an effective test of whether a person has had too much to drink.

Further experiments ruled out leg length and leg strength as factors. Souman also found no evidence to support the assumption that there is a "circling instinct" in the brain. The paths his subjects took were not big circles or spirals, but rather something more like the random squiggles a toddler makes with a crayon. At times, they looped back on themselves—the point at which walkers typically spot a familiar landmark, falsely conclude that they are *walking in a circle*, and begin to panic—but walkers almost never circled all the way back to the start. Souman concluded that on average people who are lost, without external navigational cues, will typically not travel farther than one hundred meters from their starting point, regardless of how long they walk.

A horrifying thought: On a cloudy day, in tall woods, with no other cues and no compass, a person will not travel more than the length of a football field in any one direction.

Remi and I were in just such a situation. The sky was dull pewter. Everything was covered equally in moss, so that old trick wouldn't be of any help. We hadn't brought a compass, as Johnny had advised. I did have my phone, which had a digital compass, but the one time I ever needed it, the needle spun limply, like a wandering eye. We were cut off from every form of external reference, except the trail. At one point we grew so frustrated in our circling that we struck off, bushwhacking in the direction we guessed the water must be, but we soon became nervous about getting lost, so we dutifully returned to the charmed thread.

At last, as the sky was darkening, one of us realized there were in fact two forks in the trail that looked identical, because some other hiker had previously propped up another stick against a tree in precisely the same fashion we had. Remi kicked the stick into the woods and the spell was broken. The next turn at the fork took us directly home.

+

According to Guldberg, Norwegian country folk refer to the act of walking in circles as "approaching on a false scent." The phrase— which must be somewhat rare, because the Norwegians I've spoken to have never heard it—wonderfully evokes the illusory sense of progress that attends the circling walker. Circular arguments function in much the same way: One side feels that an intellectual victory is just within reach, as does the other side. Both sides launch attacks and counterattacks and counter-counterattacks, but neither can win a decisive victory, and so the two continue "circussing round and round," like a pair of rats chasing one another around the outer rim of a barrel.

Until the latter half of the last century, ant researchers generally fell into one of two camps: they either believed ants were sentient beings, capable of learning, or they believed they were instinctive machines. As explanations based on God's role as a "prime mover" fell out of fashion in the nineteenth century and evidence of ants' problem-solving abilities continued to mount, the proponents of sentience appeared to be gaining ground. But then, beginning in the 1930s, Konrad Lorenz, the father of a new field of science called ethology, injected new life into the mechanistic argument by showing how insects rely upon "fixed action patterns"—what, in a prior era, would have simply been called "instincts"—which were genetically *coded*, rather than divinely ordained. God was swapped out, and genetics was swapped in, but the basic argument remained the same. At its heart, the debate revolved around a central paradox: If ants are intelligent, then why do individual ants behave so stupidly? But then, if ants are empty-headed, how can their colonies solve such a wide array of complex problems so brilliantly?

It was not until the advent of computing that this circle was finally broken; early computers opened up a new path forward. By programming computers to perform insect-like tasks, and by studying the (previously, overwhelmingly complex) behavior of swarms using computers, we began to understand that simple machines following a simple set of rules can ultimately make highly intelligent decisions. They are not *either* simple *or* smart; they're *both*.

Regular conferences in the new field of "cybernetics"—the study of automated systems—were held throughout the late 1940s and early 1950s, where biologists and computer scientists began discussing the considerable overlap between their fields. At the second meeting, Schneirla gave a lecture on how he had trained a common black ant to navigate a maze—being careful to regularly swap out the paper linings on the floor, so that trails couldn't accumulate—which proved that some ants could memorize basic routes. This finding suggested that individual ants had more powerful brains than many scientists had previously suggested. However, this proposition was later undercut when another attendee, Claude Shannon—famous for quantifying information into "bits"—successfully built a robotic ant, with a processor that was ten times simpler than even the most rudimentary pocket calculator. An electronic "antenna" on wheels, the robo-ant explored mazes following a simple trial-and-error program, bumping up against the walls until the antenna touched its "goal" (a button that shut off the robot's motor). On the second run-through, having memorized the maze, the robo-ant was able to complete it without touching any of the walls.

The roboticization of the insect world has continued steadily ever since. A few years ago, a researcher named Simon Garnier built robotic ants that could follow electronic pheromone trails. The trails were laid down by a row of overhead projectors, which automatically tracked each robo-ant's movement; meanwhile, light sensors were installed in each robo-ant's "head," so it could follow the other robo-ants' glowing trails. Essentially, by following just two simple rules—explore randomly until you find either a "trail" or "food," and follow the strongest trail you find—the robo-ants were ultimately able to find the shortest route through a maze.

The shift toward studying ant colonies as robotic rule-followers was mirrored by a growing sense that an ant colony could function as a single, self-organizing system, much as a computer is a collection of individual circuits. This idea was famously demonstrated in the 1970s by a Belgian researcher named Jean-Louis Deneubourg. One of his most famous experiments involved connecting a nest of Argentine ants and a food source with two different bridges. The two bridges were alike in all respects, except that one bridge was twice as long as the other. In the beginning, the ants chose between the two bridges randomly, but over time, the colony overwhelmingly chose the shorter bridge, for the simple fact that their pheromones accumulated there more quickly. The ants' system was neatly self-regulating—the shorter the path, the fresher the pheromones, and the more traffic it attracts. Here was the key: Ants may be individually stupid, but they have a high level of what Deneubourg calls "collective intelligence."

By regarding ant colonies as intelligent systems composed of individuals following simple rules, Deneubourg was able to make another leap forward: He found that he could describe their movements with mathematical formulas, which could then be used to create computer models. Ant colony algorithms—in which myriad initial routes are explored, the best ones being amplified while the others fade—have since been used to improve British telecommunications networks, to design more efficient shipping routes, to sort financial data, to better deliver supplies during disaster relief operations, and to schedule tasks in a factory. Scientists chose to model their algorithms after ants (as opposed to, say, tent caterpillars) because ants are constantly tweaking their designs and probing for new solutions; they tend to find not only the most efficient solution, but also a slew of effective backup plans.

I spoke with Deneubourg one winter morning at his home in Brussels. He greeted me at the door: a compact, spritely, gray-haired man with big ears and a wide smile. If wrinkles are a graph of all past expressions, his pointed decidedly gleeward.

From the outset of his career, when he studied under the famed systems theorist Ilya Prigogine, Deneubourg had sought to reveal the invisible systems that underlie animal behavior. He realized early on that collective intelligence extends well beyond insect colonies: indeed, historically the notion referred first to humans, and only much later to insects. The term "collective intelligence" appears as early as the 1840s, when the democracy activist Giuseppe Mazzini used it to critique Thomas Carlyle's belief that history was nothing more than the record of the actions of "Great Men." Mazzini argued that the greater goal of history was to discern the "collective thought . . . in the social organism"; for too long, he wrote, historians had focused on the petals rather than the whole flower. A fervent Catholic, he believed that the "collective intelligence" of humans ultimately stemmed from an almighty God, of whom humans were mere "instruments."

Deneubourg sought to dispense with divine explanations and instead show how collective intelligence can emerge (among insects *and* people) from the interactions of individuals. In one early paper he argued that people tend to build their settlements stigmergically, just like ants: they unconsciously modify the environment, which sends a signal instructing other people how and where to build. For example, if you build a home in an unpopulated area, other people may start to perceive that area as a nice place to build a home; build enough homes and someone may build a shop; build enough shops and someone may build a factory or a shipping port. No top-down oversight is necessary; cities can arise from the ground up.

The week before I met Deneubourg, I had talked with one of his disciples, a professor in Toulouse named Guy Theraulaz, who showed me a video of how *Messor sancta* ants dig a network of branching tunnels in a disc of dirt. Next, he showed me aerial photographs of unplanned cities—*villes spontanées*—like Benares, Goslar, and Homs. The similarity was striking. He and his colleagues had found that both these systems found a near-optimal mathematical balance between efficiency (a minimum of paths) and robustness (the good kind of redundancy, whereby the collapse of a single avenue does not lead to a systematic collapse).

"The interesting thing is that in Rome, originally that was a grid system," he said. "The whole system was destroyed by time, and then it converged into a medieval organic system." Likewise, many cities across Europe that were built on the Roman grid—Damascus, Mérida, Caerleon, Trier, Aosta, Barcelona—later collapsed back into an organic layout, as people began taking shortcuts across empty quadrants, filling in extravagant plazas, and altering the imperial road network to their needs. Left to their own devices, people unwittingly redesigned their cities precisely as ants would.

Sitting in Deneubourg's office, I thought back to this experiment. I wondered how a veteran collective intelligence researcher, knowing what he knows, would use that knowledge to design a better city. So I asked Deneubourg: If he were the mayor of a new city being built ex nihilo, like Brasília, how would he organize it?

"I would like to see the *emergence* of the town," he said. "If I was the mayor—and the probability of that happening is quite low—my attitude would be very liberal. My objective would be to offer different types of material to help the citizens find the solution that they prefer."

I found this answer somewhat surprising. By all accounts, he was an expert in the design of efficient systems. And yet he would withhold his expertise and allow the town's residents to plan their own town?

"Yes," he replied, with a look of impish mirth. "To believe that you have the solution for another person is a form of stupidity."

+

As the human population continues to swell and gather in ever more densely crowded, hive-like cities, the collective intelligence of ants begins to look all the more astonishing by comparison. Much of ants' inventiveness arises from their almost utopically (or dystopically, depending on your outlook) high degree of selflessness, which we notably lack. For example, when traversing a V-shaped ramp in a lab, army ants will construct a bridge out of their own bodies to create a shortcut across the crook. In human terms, this would be as if a businessman, while rushing off to work, decided to speed the passage of his fellow commuters by laying his body down over a gap in the sidewalk. I do not foresee us developing this kind of altruism any time in the near future. Nevertheless, there are many lessons humans can glean from the wisdom of ants.

When walking among large crowds, for example, both humans and ants naturally form lanes of traffic. However, among human crowds, those lanes break up and then slow down to reformulate every thirty seconds or so, whereas ant lanes remain in a constant, steady, orderly flow. To find out why, a crowd theorist named Mehdi Moussaid set up video cameras on balconies overlooking some of the busiest pedestrian areas in the French city of Toulouse. What he found was that a single impatient person tends to weave through the crowd, disrupting the smooth flow and slowing everyone else down. (When Moussaid told me this, I laughed in uncomfortable recognition. While commuting across Manhattan to my old job, I saw this phenomenon every morning in the crowded subway tunnel between Sixth and Seventh Avenues. Perpetually late for work, I usually was that jerk.) Ultimately, moving with the flow, rather than racing through it, gets everyone in the swarm to their various destinations more quickly.

In another surprising study on ant traffic dynamics, a former colleague of Moussaid's named Audrey Dussutour showed that ants never get stuck in traffic jams. One advantage ants have is that their highways have flexible boundaries, so they are able to effortlessly widen them as traffic increases. Even in artificially constricted conditions, however, ants still adapt better than we do. Dussutour proved this by pouring Argentine ants into a basic bottleneck-shaped maze; from above, the ants resembled a dense crowd of people trying to exit a theater through a narrow set of doors. But no matter how many ants she poured into the bottleneck, she could not induce them to grind to a halt the way people inevitably would.

She told me she had recently stumbled upon a likely explanation: She had noticed that when the crowd reaches a certain density, a small number of ants—about ten percent—will stop cold in the middle of the flow, "like stones." Remaining frozen for up to twenty minutes, the stationary ants split those moving around them into lanes, which prevents jams. By slowing down, certain self-sacrificing individuals allow the colony to move faster. This finding meshes with similar research on human crowds, which has shown that placing an obstacle like a pillar directly in front of a doorway will cleave crowds into neat rows and quicken their flow.

Talking with Dussutour, I began to envision a future where swarms of driverless cars would use ant-based algorithms to forever eradicate traffic jams. In the past, such techno-utopic schemes had always seemed far-fetched to me, because I imagined the cars would require a centralized supercomputer to coordinate their movements. (Think of the hellacious traffic jams that would ensue if that supercomputer were to malfunction.) But a growing body of research—especially in the nascent field of swarm robotics—has proven the cars could effectively coordinate themselves without a godlike hand steering them; highly sophisticated coordination can arise from the bottom up, through individual machines following simple rules.

However, Dussutour stressed that it would be a mistake to think that just because ants behave selflessly and cooperatively, they are all identical and predictable, like robots. Her work has led her to believe that the next big paradigm shift in collective intelligence research will stem from the realization that there are notable individual differences *between* members of a swarm. "People always say ants are the same," she said. "Bullshit." For example, she noted, researchers have found that fourteen percent of common black garden ants never lay a trail during their various foraging trips. Another study found that at least ten percent of foraging green-headed ants will eat whatever they find without ever bringing anything back to the nest. A third study found that as many as twenty-five percent of *Temnothorax rugatulus* do no work at all. No one knows why these selfish ants exist—whether they provide some hidden evolutionary advantage to the colony, or whether they merely demonstrate that no species is without its share of rebels and slackers.*

Systems built on universal trust are universally easy to exploit. This is why, among humans, members of utopian communes must expend an enormous amount of energy policing against shirkers and charlatans. ("Communes," wrote the social psychologist Jonathan Haidt, "can survive only to the extent that they can bind a group together, suppress self-interest, and solve the free rider problem.") Because of the premium placed on social cohesion, cooperative communities—from hives to nations—are also prone to being swayed by charismatic leaders. Experiments among shoals of golden shiner fish have shown that a single emphatic individual can alter the trajectory of an entire school of fish regardless of whether it is in the best interest of the group. Likewise, it has been found that among humans, the most confident, talkative member of a group often becomes the group's leader, more or less regardless of the quality of his or her input (a phenomenon called the "babble effect").

"The wisdom of crowds doesn't work all the time," said Simon Garnier, who runs a research laboratory at the New Jersey Institute of Technology called the Swarm Lab. "If you play it right, you can make crowds go wherever you want."

Garnier was referring to the 2004 book *The Wisdom of Crowds*, by

---

* Meanwhile, researchers have also continued to discover the extent of diversity between different species of ants. Additional varieties of trail pheromone have been discovered, including a *not-this-way* signal. And certain species of ants, it's been found, have even evolved entirely different ways of navigating, particularly in places like Arizona or Australia, where a hot, dry climate causes trail pheromones to evaporate too quickly. In these regions, ants have been found to navigate using a variety of cues, including the angle of the sun, the direction of the wind, and the texture and slope of the ground. One species of desert-dwelling ant has even been found to "count" its steps, which allows it to navigate using dead-reckoning.

James Surowiecki, which described the ways that crowds of perfectly average people can collectively make judgments that rival those of the most highly regarded experts. The canonical example of this phenomenon is an experiment run by the British scientist Francis Galton. In 1906, Galton collected data from a group of people at a country fair who were trying to guess the weight of a fat ox. Of the roughly eight hundred people who wagered a guess, most were wide of the mark. However, the average of all their guesses was nearly perfect.

This experiment would later be repeated many times. Oddly, researchers learned that the key to the experiment was that each person needed to judge the weight of the ox independently, without sharing their guesses with one another. In similar experiments where people were given access to one another's answers, the collective intelligence of the group worsened. Often, the early guesses provoked a false consensus to form, a vicious circle that caused the later guesses to hurtle toward ever-greater error. "The more influence a group's members exert on each other," wrote Surowiecki, "the less likely it is that the group's decisions will be wise ones."

I was startled when I first read this finding, because it appeared to contradict everything scientists have learned over the past three hundred years about how trails form. When trails are taking shape, every member of a crowd has access to every previous walker's guess, because their choices are written right there in the dirt. And yet trails nevertheless tend to find optimal routes across the landscape, rather than veering off on wild, mistaken detours. How could this be?

I recently ran across an answer in a paper by a bioscientist named Andrew J. King, who conducted a clever update on the famous Galton ox-weighing experiment. In it, he asked a group of 429 people to guess the number of sweets in a jar. But this time, he made a few tweaks: He gave each member of one group access to the guess of the previous guesser. He gave another group the mean of all previous guesses. And he gave the members of a third group access to a random previous guess. All these pieces of additional information, as predicted, skewed the group's answers for the worse. However, when he gave members of a fourth group access to the "current best guess"—the previous guess which was currently closest to the mark—he found that that group not only outperformed the three others, but in certain respects it also outperformed the classic Galtonian, private-information-only group.* Among crowds, sharing more pieces of random information is generally unhelpful—like rumors swirling through a school, they amplify toward ever-greater falsehood as they go. But more *reliable* information—even if it is not perfectly correct—kicks off a process of fine-tuning, until the answer is revealed.

Every trail is, in essence, a best guess: An ant does not leave a strong pheromone trail unless it has found food, which means that it has already made a correct calculation of where the food is. The same rule applies to humans—we generally don't make trails unless there is something on the other end worth reaching. It's only once an initial best guess is made, and others follow it, that a trace begins to evolve into a trail.

As Huxley argued, the same pattern underlies all scientific progress; best guesses are ventured, which, over time, become better guesses. Thus a trail grows—a hunch is strengthened to a claim, a claim splits into a dialogue, a dialogue frays into a debate, a debate swells into a chorus, and a chorus rises, full, now, of clashes and echoes and weird new harmonies, with each new voice calling out:

This way . . .

This way . . .

This way . . .

---

* The private-information-only group was closer in regards to the median, but the current-best-guess group was closer in regards to its mean. On the whole, both were much closer than the other three groups.

CHAPTER 3

S TRANGE CREATURES paced the pale autumn grass. Out the left window of our Land Rover stood a trio of wildebeests. Out the right: a muscled cluster of elands. The truck slowed to a crawl. Up ahead ambled a flock of mountain goats, clotting the road. As we crept forward, a nearby giraffe craned down and peered through our windows with the slow, sleepy eyes of a courtesan.

Beside me, in the driver's seat, was Nidhi Dharithreesan, a biologist specializing in the herd behavior of large mammals. She pointed out the more obscure species—gnu, oryx, kudu, addax, waterbuck—whose names reminded me of characters in a science fiction novel. Over the months she had spent watching these animals, she had formed lovingly frank opinions about each: White rhinos are "sweet," but zebras are "assholes." Elephants will "tear up everything" if given the chance. Male kudus, like boys at their first school dance, are more interested in performing elaborate mating displays than in actually mating.

I had met Dharithreesan at the Swarm Lab in Newark, where

she was finishing her PhD. Alongside her insect-obsessed colleagues, her interest in swarms of giant furry beasts was somewhat unusual. However, she informed me that there was a great deal of overlap between their work and hers: Mammals, like insects, aggregate on a massive scale, share information, and create highly efficient networks of trails. If you were to step into a hot-air balloon and float high above the Serengeti during the annual great wildebeest migration, the herds of ungulates would resemble nothing so much as an invasion of safari ants.

By studying insects, I had learned that trails can function as a form of external memory and collective intelligence. Bugs benefit from trails because they are tiny and small-brained but nevertheless must manage huge, complex tasks. But why, I wondered, do we big-brained, highly individualized land mammals—the greatest class of walkers in the known universe—feel the need to trail after one another? Why not walk alone, utterly free?

It can be tough to puzzle out why other animals do what they do. Between humans and the rest of the animal kingdom lie near-insurmountable psychic and linguistic barriers. And yet humans have always been peculiarly curious about what motivates other animals, and whether those motivations resemble our own. I have yet to find a culture on Earth that does not speculate about the interior lives of our animal brethren. And for good reason: our survival often depends on it. To understand the psychology of their prey, many indigenous hunting societies perform magical rites, including ritual trances, sacrifices, ceremonial dances, various forms of fasting, even self-mutilation. Driven by that same basic question, Western scientists perform elaborate experiments and concoct dazzlingly complex computer models. The understanding we've sought and the bonds we've formed with other animals have, over millions of years, made humans—all of us, from the gazelle hunters of the Kalahari to the cat fanciers of Tokyo—into who we are today.

Traditionally, humans have learned to empathize with other species in three distinct ways. Perhaps most commonly, we have bonded with animals by living alongside them: housing them, feeding them, breeding them, and herding them from place to place, until we attained a kind of loose symbiosis. Conversely, we have also learned about them by hunting and killing them, which produces a wholly different kind of mind-meld—the cold empathy of the predator. And most recently, we have begun studying them: cataloguing what they eat, tracking where they travel, testing how they react, and modeling how they organize themselves.

This is how I found myself sitting in a Land Rover beside Dharithreesan. I had decided to spend some time trying out each of these oldest forms of cross-species communication: watching, herding, and hunting. (Naturally, I began with the one that intimidated me least.)

In all three of these pursuits, I would learn that trails provide a helpful (if narrow) portal into the minds of other animals. Despite what we sometimes imagine, the animal world is not a rigidly compartmentalized place, like a child's coloring book, with zebras on one page, giraffes on another, and lions on the next. Animals are all intermixed, interdependent. Flocking and stalking, they follow one another across landscapes. And in those places where their most vital needs overlap, trails inevitably appear.

Human animals are not excluded from this collaborative process; the earliest humans no doubt relied on the paths of other land mammals, just as many modern roads overlay old game trails. By following in the footsteps of other animals, we have learned to intuit their intentions. Expert trackers, it is said, begin to identify themselves with their quarry; this inclination allows them to follow trails that intermittently vanish, and even to experience the same sensations the animals felt as they walked—the prick of a thorn in a paw, the soft give of warm sand under hoof—a process that is sometimes referred to as "becoming the animal."

We may not be able to read other animals' minds, but we can read their trails. In learning to do so, we have become recognizably human: hunting animals, we sharpened our intelligence and invented some of our earliest technologies; herding animals, we reaped the reliable luxuries of milk and meat, leather and wool; harnessing animals, we tilled fields, transported goods, and built cities; and studying animals' wisdom, we have increased our own. That long, slow waltz across continents—as humans and animals clashed, meshed, and, ultimately, began to prop one another up—would in time transform us all.

PART I

Watching

A sugary rain began to sift down as Dharithreesan and I watched the animals enact the mundane chores of being alive. An addax delicately scratched an itch on his back with one long curved horn. A baby antelope wobbled beside its mother, who bent down, stuck her nose up under the calf's hind legs, and licked its rear. The calf looked over at us, blissfully unabashed.

Staring at these herds of striped and spotted ungulates, I could almost be fooled into believing that we were on a safari in some far-off veldt, if it weren't for certain discordant details: the high steel fences, the cartoonish faux-wood sign reading AFRIKKA, and, most jarring of all, in the distance, the swooping steel scribbles of roller coasters. In fact, we were in an enormous outdoor zoo— reportedly "the largest drive-through safari outside of Africa"— attached to the Six Flags Great Adventure theme park in suburban New Jersey, less than a two-hour drive down the turnpike from New York City. The safari park was introduced in 1974, alongside attractions like the world's biggest hot-air balloon and the world's largest teepee. It now contains over twelve hundred animals from six continents, including a sizable population of African herd animals.

Dharithreesan set up a camera on a tripod on her windowsill to film the animals' movements. She began jotting notes in a field journal: date, time, temperature, weather conditions, and any notable behavior. She was in the preliminary stages of a multiyear campaign to tag the park's African ungulates with GPS-enabled collars. The data would be transmitted wirelessly to a receiving station then relayed to the Swarm Lab, where it would eventually help solve the riddle of why mammals form herds.

One of the most prominent explanations, which she hoped to test, was called the "many eyes theory." The more eyes a herd has, this theory holds, the more likely it is to detect a predator or a new source of food. By taking turns scanning the plains, more herd members are free to graze in peace. Many African ungulates—zebras, wildebeest, gazelles, antelopes—tend to live in mixed herds, perhaps because the strengths of one species make up for the deficiencies of another. Zebras, for example, are nearsighted, but have excellent hearing, while giraffes and wildebeest have keen long-range vision. By herding together, they increase their chance of spotting (or hearing) the approach of a stalking lion.

Dharithreesan planned to test this theory by installing electronic collars on all the ungulates, which would track not only each animal's location, using GPS, but also employ gyroscopes and accelerometers to record which direction its head was pointing. Scientists have so far conducted only a few studies like this on the dynamics of mixed-species herds. The logistics were staggering, Dharithreesan told me. "You can't really do this type of study in the wild, because there's just too much space; we don't have the resources," she said. "And you can't quite do it in a laboratory setting, because these animals are *huge*." Fortunately, the owners of Six Flags Great Adventure had unwittingly built the ideal scientific testing ground.

Out in the wild, scientists often sacrifice this kind of granular data for a much broader scope. With the rise of satellite technology, humans have suddenly acquired a god's-eye view of how animals move across vast stretches of land. Previously, to track a group of animals in the wild, scientists had to tag them with radio collars and then, using jeeps equipped with special antennas, chase after the tagged animals. Now with GPS collars, researchers can tag an animal, let it roam for months, and then download the collar's data either manually or, increasingly, wirelessly. This new technology—paired with ever more detailed satellite imagery—is revealing how groups of mammals create and pass down migration routes from generation to generation. Some of the oldest of these migratory routes, like those of Canadian mountain sheep, likely stretch back tens of thousands of years.

A few years ago, an ecologist named Hattie Bartlam-Brooks attached GPS collars to a group of zebras in Botswana's Okavango Delta to track their grazing patterns. At the time, it was widely believed that the zebras never left the delta, so when a large number of the zebras disappeared from sight at the onset of the rainy season, Bartlam-Brooks assumed they had been eaten by lions. Then, six months later, the tagged zebras reappeared. When Bartlam-Brooks recovered their collars and downloaded the data, she discovered that the zebras had somehow walked halfway across the country, to feed on the sprouting grasses of the Makgadikgadi salt pan.

By reading through old hunters' and explorers' records, she learned that a large zebra migration had once existed along that same route, but it had been severed when the Botswanan government installed hundreds of miles of veterinary cordon fences in 1968. One of these fences blocked the zebras' migratory route for decades before the government finally dismantled it in 2004. Since the fence stood for thirty-six years, and the average lifespan of a zebra is only twelve years, no living zebras could have possibly remembered making that trip. But then, I wondered, how could the zebras have known where to go?

When I spoke to Bartlam-Brooks on a long-distance call to Botswana, she quickly ruled out my first guess: there was no grassy runway—as I had imagined—that lured them across the country. Instead, they had to pass over hundreds of miles of dry Kalahari scrub. The study's coauthor, Pieter Beck, explained that migrations, by definition, involve not only long distances, but also high stakes: in a migration, there is always a considerable "energetic cost" to the journey. Every voyage is a gamble. (This may explain why not all the zebras ended up taking the trip. Even among zebras, there are bold and timid individuals.)

Because the cost of unsuccessful exploration is so high, successful migration routes are precious and hard-won. Older herd members teach the routes to their children, passing them down as a kind of traditional knowledge. But like all traditions, migratory routes are delicate. Once a route is disrupted, it rarely reemerges. What Bartlam-Brooks had apparently uncovered was a rare instance of a species reviving their ancestral lifeway.

But still I wondered: How? I pushed Bartlam-Brooks to venture a guess.

Her answer surprised me. She said that her hunch was that, through a series of exploratory walks, the zebras might have followed a chain of elephant trails that led them from water source to water source all the way to the salt flats.

"Elephants are obviously much more long-lived than zebras," she said, "so when the fence went down, it's very possible that some elephants remembered that old historical pathway that they used to take. Elephants could have easily re-created game trails, and zebras may well have just followed them."

Of course, I thought. *Elephants.*

+

I once spent three weeks hiking across the grasslands of Tanzania, through the Ngorongoro Crater, to reach an active volcano called Ol Doinyo Lengai. During the day, we would occasionally spot surreal animals grazing in the distance: giraffes, buffalos, a dozen kinds of antelope, their horns twisting upward like Chihuly glass. At night, hyenas rubbed up against the walls of our tents, giggling menacingly, their bitter musk seeping through the nylon.

The walking was hard. The land was covered with tall yellow grass and corrugated with deep trenches, called drainages. Fortunately, elephants had created a convenient system of trails for us to follow. They proved remarkably clever route finders. On many occasions, after following an elephant path down yet another steep drainage, I marveled at the fact that the elephants had somehow selected the shallowest gradient available for at least a hundred yards in either direction. I wondered: *How does the elephant know where to go, when even we, with our maps and compasses, do not?*

Descriptions of the elephant's topographic genius—a cherished gift when the land is barbed with thorn bushes and aflame with stinging nettles—lie sprinkled throughout colonial literature. "The sagacity which they display in 'laying out roads' is almost incredible," wrote Sir James Emerson Tennent about the Ceylonese elephant. "The elephants invariably select the line of march which communicates most judiciously with the opposite point, by means of the safest ford." The same is true of African elephants, wrote the poet Thomas Pringle. Their trails always seemed to have been cut "with great judgement, always taking the best and shortest cut to the next open savannah, or ford of the river; and in this way they were of the greatest use to us, by pioneering our route through a most difficult and intricate country."

The trail networks of elephants can often cover hundreds of miles, connecting distant food sources and salt licks, while expertly avoiding whatever obstacles—canyons, mountains, dense forests—might stand in their way. But how did elephants figure out exactly where their trails should go? What allowed them to find the far-off mineral deposits they needed or the shallowest ford across a river?

I went in search of an answer to this question. And fortunately, I knew just where to start. My old AT thru-hiking buddy "Snuggles" (real name: Kelly Costanzo) happened to work at a place in rural Tennessee called The Elephant Sanctuary. There, nineteen female elephants, which had formerly been held captive in zoos, circuses, and backyards, now roam free across 2,700 acres of open forest.

One summer afternoon I traveled to the sanctuary to pay Kelly and the elephants a visit. I drove down a dirt road and then pulled up to a password-encoded steel gate, which bore a sign that read WARNING: BIOHAZARD. The whole property was surrounded by two rows of tall steel fences, one topped with barbed wire. At first glance, the place had the feeling not of a sanctuary, in the holy sense, but of a top-secret compound where monsters are made.

Kelly met me on the other side of the gate in her car and led me inside. She showed me to her home, a sunny, one-story ranch house on the sanctuary property. She shared it with two dogs, four cats, and one ingenious gray parrot (who had learned to bark, meow, trumpet like an elephant, and chime like a cell phone). From her backyard, off in the distance an elephant with a crippled trunk could be seen ambling along the fence line. This juxtaposition between the warmly familiar and the eerily exotic was one I would experience often during my stay at the sanctuary.

Kelly and I sat up late that night, drinking beer and reminiscing about the AT. She recalled one darkly comical morning we'd spent together in Erwin, Tennessee: We were sitting at a greasy lunch counter, eating breakfast before hitching back to the trail, when a local man turned to Kelly and began recounting, with evident pride, Erwin's claim to fame—the fact that in 1916 they had "lynched" a mad elephant, named Mary, before a crowd of thousands. He gestured to a framed black-and-white photo on the wall immortalizing the event. The man—having no way of knowing that Kelly had been working intimately with elephants for years—waited for her reaction, expecting, perhaps, coos of intrigue. The best she could manage was to stare back at him wordlessly, her face fixed in open horror.

The next morning she gave me a tour of the grounds. We began at the first "barn," where the African elephants were housed on cold days: a vast, echoey shed with diaphanous polycarbonate walls, heated floors, and a corrugated steel roof. The elephants' stalls led out onto a cropped lawn and, beyond it, a vast forest. Outside, the elephants' turf was surrounded by a fence made up of rectangular steel arches, like a row of gigantic steel staples, that were large enough for a human to pass through, but too narrow for elephants. I asked if we could walk through to the other side. Kelly gravely shook her head.

On the other side of the fence, twenty or thirty yards away, hulked a giant creature the color of unglazed Japanese pottery. We drew closer, but not too close. The giant had two nubby white tusks and a trunk like a crocodile's gray tail. Her name, Kelly said, was Flora.

Flora's trunk reached out to sniff us, slithering bonelessly over the fence. That one organ, I had read, could pluck a blueberry, uproot a tree, jet gallons of water, or catch a scent from miles away.

"What an amazing animal," I said.

"I know, they're really amazing," Kelly sighed. "But she's one that would, like, kill somebody if she had the option."

"So if I were to walk up to the fence . . ." I trailed off.

"Oh gosh. She'd probably swing her head over it and try to grab you and kill you."

In the two decades since it first opened, the sanctuary has suffered only one deadly incident, in 2006, when a caregiver was stomped to death by an elephant named Winkie. In the aftermath of that event, the caregivers pared back their contact with the elephants; no more are the days when people could stroll through the fence to pat their favorite elephants on the trunk. Kelly suspected that many captive elephants grow unusually aggressive toward humans as a result of the trauma they suffer in captivity. "There are some that just have such deep, deep scars that they'll never trust humans," she said.

Flora, for example, had almost certainly witnessed the slaughter of her parents as a young calf. (Many scientists now believe that elephants understand the concept of death; elephants have even been seen grieving over the gravesites of family members.) Flora was subsequently abducted, put in chains, sent overseas, broken by trainers, and forced to perform for the amusement of paying customers. In the circus where she had performed, she was billed as "the world's smallest and youngest performing elephant."

Flora bent back her trunk until it touched her forehead, forming a bubble-letter *S*. The inside of her mouth was shell pink. It curled softly in on itself like the bloom of a snapdragon. "She's so pretty," Kelly said. "Look at that mouth!"

We stood and stared until Flora lost interest in us and sauntered off across the yard and into the trees. "This whole area used to be pines when I first came here," Kelly remarked. "The elephants just knocked them all down."

I asked her why.

"They're savannah makers," she shrugged. Many experts believe elephants act as what biologists call "ecosystem engineers." Research by zoologist Anthony Sinclair has shown that elephants take advantage of wildfires to clear patches of forest and convert it to grassland. The elephants let the fire do the hard work of burning down most of the trees, and then they pluck out the tender green shoots that spring up in the wake of the fire, to prevent the trees from growing back. Along their trails, elephants have also been known to "garden"—pulling up saplings of encroaching trees and dispersing the seeds of the fruit they eat. In dense jungles, where wind cannot disperse seeds across long distances, elephants play Johnny Appleseed; their dung spreads the pits of large fruits like mangos, durian, and (fittingly enough) so-called elephant apples. As a result, the trails of elephants are often lined, conveniently, with their favorite fruit trees.

A young caregiver named Cody shuffled over, wearing a frayed baseball cap and a T-shirt emblazoned with skulls. I asked him something I had been wondering since I arrived there: Do the sanctuary's elephants make trails, as they would in the wild?

I half expected him to say no. It appeared to me that whatever reasons elephants might have for following trails in the wild were absent at the sanctuary: they did not need trails to help them navigate long distances; there are no swift rivers to cross, no mountains to climb. However, Cody and Kelly both nodded; the elephants seemed to love making trails, they said. Cody pointed out a faint elephant trail that ran across the yard and along the fence line. There were others; narrow two-lane tracks that crisscrossed all over the property. Neither of them knew where most of the trails came from; most had simply appeared years ago. One elephant, named Shirley, had created a trail that led to the grave of her former companion, Bunny. (Today, the caretakers call it "Bunny's Trail.") They said that some of the elephants would stick to certain trails even if those trails did not provide the fastest way to get from one point to another.

I asked Kelly why she thought they were so trail obsessed, even now, after a lifetime of captivity, when trails weren't necessary for their survival.

She smiled and shook her head.

"I'm guessing it's deep-rooted," she said.

+

Later in the day I began piecing together more clues. Kelly and I were visiting the Asian barn, on the other side of the property. Outside the barn sat two dust-yellow elephants named Misty and Dulary. They were smaller and pudgier than the Africans. Misty was lying on her side on the ground, while Dulary stood guard. Spying us, Dulary walked slowly over to the edge of the fence and stared. The shape of her forehead resembled a bull's skull: a pair of bulging orbits hourglassing into deep, hollow temples. Her trunk hung like the hose of an old gas mask. Where the whites of her eyes should have been, there was black.

Misty rolled over onto her stomach, tucked her knees underneath her, and, in a toddlerish motion, stood up, front legs first. Her face was noticeably chubbier and wrinklier than her companion's. Kelly described it as "smushy," like a marshmallow. Misty walked over to Dulary. The two stood side by side: the embodiment of cuteness, the visage of death. They began feeling each other with their trunks, sweetly. Then, as if on cue, they both pissed a torrent.

Cody soon stopped by on his rounds to check in on Misty, which set in motion a smoothly choreographed routine. He walked up to the edge of the fence. Misty turned around and stuck out her knobby tail. He pulled on it gently. Then she lifted her foot. He gave it a hug.

Kelly told me that the caregivers trained the elephants to lift their feet so that they could provide medical treatment. The caregivers dedicated hours each week solely to mending damaged feet— cracked toenails, abscesses, infections—which were common among elephants that, in their former life in captivity, once spent much of the day standing on hard concrete. Making matters worse, a few of the elephants, during their former lives in captivity, had developed odd tics—some rhythmically swayed their bodies from side to side; others tossed their trunks forward and back—which zoologists call "stereotypic behaviors." Elephants are magnificent long-distance walkers; in the wild, they can travel up to fifty miles per day. So when they are confined, they will often begin fidgeting to release the excess energy. Because the movement releases endorphins, it can become ingrained as a form of self-soothing, which over time can lead to joint and foot injuries. Foot problems, Kelly pointed out, were the leading cause of death among captive elephants.

Despite its stump-like appearance, an elephant's foot is an oddly delicate appendage. Hidden within that fatty cylinder lies a bone structure resembling a kitten-heeled shoe. This tiptoed design allows elephants to be surprisingly nimble climbers; one hunter in colonial Africa described finding elephant trails leading up the face of a cliff he considered "inaccessible to any animal but a baboon." In the circus, elephants have even been trained to walk tightropes.

On flat ground, a disc of fat on the sole of each foot dampens much of the impact of walking—a nine-thousand-pound elephant exerts less than nine pounds per square inch of pressure underfoot— giving them a soft, quiet tread. Their tendency to clear paths further facilitates silent creeping. Dan Wylie, the author of a cultural history of the elephant, recounts the story of a group of rangers in the Zambezi Valley of Zimbabwe who, while camping out one night, unwittingly fell asleep in the middle of an elephant path. In the morning, they realized that an elephant had walked directly over their bodies without waking even a single one of them. Its footprint was stamped into the groundsheet between where they lay.

Yet more incredibly, it appears that elephants may also use their feet to listen for messages from distant members of their herd. An elephant researcher named Caitlin O'Connell-Rodwell—who previously studied the way Hawaiian planthoppers communicate by sending vibrations through a blade of grass—found that elephants can use what she called their "giant stethoscope feet" to detect distant alarm calls transmitted through the ground. She theorized that elephants' feet could also detect the rumble of thunder from up to a hundred miles away. If so, this would help explain their mysterious ability to travel across vast distances to the precise location of newly rain-fed land.

The feet, I realized, might provide a clue to how elephants can find the easiest route across hundreds of miles of jungle or desert. In fact, when I thought about it, the whole of an elephant's body is perfectly engineered for creating trails. With their powerful sense of smell and hearing, elephants can detect food, water, and other elephants from many miles away. With their broad shoulders, they can bash through dense brush. Because of their immense weight—it requires twenty-five times the amount of energy for an elephant to climb one vertical meter as it does to travel the same distance on flat terrain— elephants will travel to great lengths searching for shallow inclines. (This explains why, as I had noticed in Tanzania, elephants always find the easiest place to cross a river drainage.) Elephant brains, too, are ideally tooled for trail-making; their fabled memory is no myth, particularly in regards to spatial information. They have evolved, it seems, to learn the land.

On top of everything else, the family structure of elephants is extremely conducive to trail creation. Typically, herds of female elephants travel single file, with their matriarch in the lead. It is the role of each matriarch to memorize the location of grazing spots and watering holes; over repeated journeys, those routes are taught to the younger elephants, one of which will grow up to become the next matriarch.[*] This hierarchical, clan-based form of travel likely dates back to the dawn of their species. Paleontologists have discovered a six-million-

---

[*] Kelly noted that when formerly captive elephants come to the sanctuary and are reconnected with other female elephants, for some reason they never re-form large herds with a distinct matriarch. Some paired off like Misty and Dulary; others remained solitary. One elephant named Tarra famously befriended a golden retriever named Bella.

year-old "Proboscidean trackway"—the fossilized footprints of thirteen female elephant-like creatures moving together along the same
path. Over time, given their sheer size and social structure, elephants
will—whether they want to or not—print out trails in their passing.

The unique physiology and social structure of elephants explains
how they create such elegant trails. But *why* they follow them remained unclear to me. With all these powerful instruments of perception at their disposal, do elephants need trails, or are trails merely
a by-product of walking? Do they give them any more thought than
we give the footprints we leave in an inch of newly fallen snow?

I asked these questions to an ecologist named Stephen Blake,
whose work focuses on how animal movement affects the land's ecology. In the late 1990s, Blake began studying the ways that forest
elephants disperse fruit seeds throughout Nouabalé-Ndoki National
Park in northern Congo. To better understand where the elephants
were traveling, he started creating a rough map of their trails. Hiking
through the swamps and rainforests, he systematically surveyed the
trees surrounding the elephant paths, paying special attention to trail
intersections. He discovered that the trails overwhelmingly found
their way to clumps of fruit trees or to mineral deposits. Other studies
have revealed that in desert landscapes, where elephants are forced to
cover vast distances to survive, trails likewise tend to connect watering
holes and grazing areas. "Lo and behold," Blake said, "just like all the
footpaths in England lead to either the pub or the church, all elephant
trails tend to lead to something that elephants want to get to."

Blake suggested that, even if these trails are not created with any
conscious foresight, they quickly come to serve a number of functions. "Let's say you've got naive elephants in a forest they've never
been to, and there are various trees dotting the area," he said. "I suspect at first elephants would go bumbling around and wait until they
bumped into a fruit tree, and in doing so, they would have knocked
down a load of vegetation to get there, and they would perhaps re-

member the geographic location of that tree. And even if they didn't, as they kept bumbling, they're probably going to create a path of least resistance through otherwise thick vegetation. So, just like when you go for a walk through the woods, you get to know which trails lead to a good place and which don't, presumably elephants learn, and certain trails start being reinforced."

In the jungle, where resources like fruit trees are plentiful but randomly distributed, or in the desert, where resources like watering holes are rare and far-flung, trails serve to reduce the amount of bumbling (a costly activity, energy-wise) and lower the chances of an elephant missing the mark. "Elephants, just like people, get disoriented," Blake explained. Trails can reorient lost elephants and reconnect disparate populations. In doing so, trails serve as "a form of societal spatial memory"—a collective, externalized mnemonic system, not unlike that of ants or caterpillars.

And it turns out that the big brains and powerful sense organs of elephants, rather than obviating the need for trails, in fact allow them to create vaster and more complex trail networks. Instead of just signifying *this way leads to something good,* as a caterpillar's trails do, memory allows for the nuance of new categories: Animals can learn that *this way leads to fruit, this way leads to water,* and even (in an elephant's case) *this way leads to my sister's grave.* Animals can begin to feel oriented within a network, aware of where they are in relation to the things they need. In a sense, memory can serve as a trail guide—not necessarily a full record of where things are, but an index of how to quickly access them.

Given their powerful memories and the similarity of their trail networks to our own, I wondered if elephants had grown to regard trails roughly the way we do—which is to say, symbolically. I asked Blake: Does an elephant know what a trail *means*? In other words, do elephants recognize a trail not just as an easy place to walk or an instinctive attraction, but as a symbolic indication that something worth reaching lies at the other end?

I had posed the same question to dozens of animal researchers—with areas of expertise ranging from caterpillars to cattle—without ever getting a satisfying answer. Blake's response was unequivocal: "For sure."

Creating symbolic trails may seem an onerous mental feat for a nonhuman animal, but in fact, with a vast enough plot of land to memorize, thinking symbolically becomes the path of least resistance. It collapses a complex environment down into neat, easily recognizable lines, and then individuates each of those lines according to its destination, like the color-coded lines of a subway system. Certainly, animals *could* navigate without them, but it would be more difficult, and natural selection, as Richard Dawkins has observed, "abhors waste."

Relying on trail networks for survival is not without its dangers, though. In places like the Congo, the elephant trail network has recently been disrupted by logging operations, which has left the elephants dangerously disoriented. Blake described the effects of the destruction this way: "Let's say you take a vibrant city that was bombed to buggery in World War II—you take Coventry or Dresden—that had transport networks all over the place. It was interconnected, everybody knew how to get from one side of the city to another. All of that infrastructure that people understood, it was the basis of their lives—and then it got the crap bombed out of it, and buildings fell, piles of rubble everywhere. Then, you just have chaos. Similarly, when you selectively cut a rainforest—you send in bulldozers, you chop out one or two trees per hectare, you pull them out, and in doing so you knock down a lot of other trees, you create other roadways—you just *erase* what was there. Even if you don't go in and kill the elephants, you've done astonishing damage to that sort of beautiful latticework, that functioning system."

Once a trail system or a learned migration route is severed—as, increasingly and alarmingly, they are, due to human habitation and industry—it rarely reestablishes itself, and the population suffers crippling losses. This is why the zebra migration route Bartlam-

Brooks had uncovered in the Okavango served as such a startling, hopeful discovery. If her theory is correct, it means that, once the fence fell, a herd of elephants managed to revive one of their ancestral routes, which led to a blooming valley of fresh grass hundreds of miles away. Once the elephants' path was reestablished, hordes of other animals could then benefit from the wisdom revealed by the passage of those broad, sensile feet.

PART II

Herding

After I left the elephant sanctuary, I kept thinking back to the image of Misty obediently raising her hind foot so that Cody could inspect it. Before I arrived there, I had expected the elephants would be distant and aloof, carefully avoiding humans, the spindly creatures who had once terrorized them so viciously. But watching Misty and Cody, what struck me about their exchange was that it seemed so calm, so natural; it had none of the air of begrudging acquiescence one often sees when elephants are forced to perform silly tricks. It was gentle, almost affectionate. What it reminded me of most, I later realized, was a handshake.

Knowing how much violence can go into training circus elephants, I was curious about how much coercion had gone into teaching Misty this gesture. According to Kelly, the process was totally pain-free. It was a textbook case of Pavlovian conditioning: First, the caregiver teaches the elephant to associate the sound of a clicking device with a treat, like an apple. The purpose of the clicking device— called a "bridge"—is to let the elephant know the exact moment it has completed the desired behavior. The caregiver begins by clicking and giving the elephant a treat.

Click: treat.

Click: treat.

Click: treat.

The caregiver does this until the elephant reaches out her trunk for a treat whenever she hears a click.

Then, the caregiver touches the elephant's leg with a stick (or "target pole").

Stick: click: treat.

Stick: click: treat.

Stick: click: treat.

Finally, the caregiver holds the stick a few inches away from the elephant's leg and says "foot." Then, the caregiver waits.

"Foot."

"Foot."

If and when the elephant finally lifts her leg to touch the stick, *click: treat.*

Using a roughly similar method, Kelly was training some of the elephants to receive treatment for their tuberculosis. The elephants needed a course of medication, but they had refused to swallow the foul-tasting pills. So once a day, seven days a week, each infected elephant had been trained to wait patiently while Kelly or another caregiver inserted her rubber-clad arm into the elephant's rectum. According to Kelly, the elephants did not enjoy receiving this treatment any more than she enjoyed administering it.[*] In the long coevolution of humans and elephants on this planet, this is where we have ended up. First, we ran from them, then we hunted them, then we enslaved them; and now we—some of us, at least—do disgusting things to keep them alive.

The elephants fortunate enough to have found their way to the sanctuary probably enjoy better lives than any other elephant in North America—roaming freely across many acres of open forest, well fed, free from predation, their every scab and sneeze worried

---

[*] Fortunately for everyone involved, in the years since I last visited the sanctuary, the caregivers have devised a way to mix the medication into the elephants' food.

over. Nevertheless, I imagine they must sometimes feel like Vonnegut's Billy Pilgrim, comfortably trapped on an alien planet—probed, pampered, and constantly, if politely, surveilled. Despite the caregivers' best efforts to simulate their natural environment, simply by virtue of being cut off from their families and their homeland, the elephants have been made strange to themselves.

Elephants have been tamed countless times throughout history, but have never been domesticated. Nearly every trained elephant—from Hannibal's war beasts to Barnum's ballerinas—was wild born, and subsequently "broken," as animal trainers used to say. This is what separates a tame animal from a domesticated one. A domesticated animal, like a sheep or a cow, never needs to be broken, because it has already been bred to live comfortably in a human environment. We have sculpted it, right down to its genes, to fit into our version of the world.

In *Guns, Germs, and Steel,* Jared Diamond noted that the "Major Five" domesticated animals—sheep, goats, cows, pigs, and horses—share a rare set of just-so features: they are neither too large nor too small; neither too aggressive nor too fearful; they grow quickly; they can rest and reproduce in close quarters; and they abide by what Diamond has called a "follow-the-leader" social hierarchy. Channeling Tolstoy, he quipped: "Domesticable animals are all alike: every undomesticable animal is undomesticable in its own way."

Elephants share some of these traits (a strict dominance hierarchy), but not others (they are too big, too restless, and grow too slowly). As almost-domesticates, they have been entered into a rather grisly lottery: each year, for the past four and a half millennia, an unlucky few are abducted, broken, and forced to work for humans, while the rest roam free.

In his controversial 1992 book *The Covenant of the Wild: Why Animals Chose Domestication,* the science journalist Stephen Budiansky argues that "virtually all of the important characteristics that set apart domesticated animals from their wild progenitors" can be accounted for by a single biological phenomenon called "neoteny": the retention of juvenile traits in an adult animal or, as Budiansky playfully puts it, "perpetual adolescence." These traits include notably cuter physical features and more flexible brains. In stark contrast to the behavior of adults, which tends to be rigid, neotenates explore, play, and solicit care just like the young. Notably—and crucially—they also tend to lack a defensive or fearful posture toward other species and new situations. These traits are most notable in dogs (which were, not coincidentally, the first animal to be domesticated), but are also evident, to varying degrees, in all of the Major Five domesticates.

Yet more striking is Budiansky's panoramic description of how humans and domesticated animals, having locked themselves into a symbiotic blood pact, proceeded to colonize the earth. What unites humans and our motley alliance of herd animals, he suggested, is that we are all "edge-dwellers," opportunists who continually exploit new and shifting landscapes. Our flexibility is our chief weapon; we are "the scavenger or grazer that can eat a hundred different foods, not the panda exquisitely adapted to living off nothing but huge quantities of bamboo." Far from having been enslaved, domesticates "chose" (read: evolved) to rely on humans and the changes we wrought on the land-scape. The Major Five animals—along with chickens, guinea pigs, ducks, rabbits, camels, llamas, alpaca, donkeys, reindeer, exotic bovids like the yak, and a handful of other species—became domesticated for the same reason some people chose to give up the free-ranging life of a hunter-gatherer to toil as agriculturalists: because it allowed them to outbreed and outcompete their rivals. It was easier to follow the shepherd into the pen than to strike off alone into the wilderness.

It is no coincidence that domestic dogs, sheep, goats, horses, and cattle all vastly outnumber their wild counterparts. Meanwhile, farming and raising animals has allowed one hundred times more people to live on the same area of land as hunting and gathering. While many animal rights activists argue that animal husbandry is unnatural and cruel, Budiansky vigorously defends the pastoral lifestyle. "In raising animals," Budiansky writes, "we are reenacting something not as old, culturally speaking, as hunting, but in a way more profound, for the rise of animal agriculture is an example of evolution operating at its highest level—on *systems* of species, one of which is us." Together, we agro-pastoralists, our livestock, and our crops reshaped ourselves to suit one another's needs. In doing so, we evolved into an indomitable (if not infallible) ecological system that has reshaped the earth.

+

A trail forms when a group of individuals unites to reach a common end. Many of the animal world's most impressive trails therefore come from herds of big mammals—elephants, bison, a varied assortment of African ungulates—which are able to expertly band together.

For all our lofty scientific studies, though, we still have only a vague sense of how herds operate. As I began to think more about the dynamics of a herd, it occurred to me that humans have been intimately studying one herd animal, up close, for millennia: the humble sheep. To observe how sheep collaborate to form trails—and, moreover, how humans and sheep collaborate to change landscapes—I resolved to try my hand at shepherding.

At the heart of every sheep lies an inherent tension between obedience and disorder. Every child knows that sheep are archetypal herd animals; indeed, the word *sheep* is virtually synonymous with something that blindly follows those around it. This trait led Aristotle to deem sheep "the most silly and foolish animals in the world." And yet, in the weeks I spent working as a shepherd in the spring of 2014, I learned that the better one gets to know sheep, the less sheep-like they appear. In fact, each individual sheep has its own personality and temperament. Some are stubborn and (relatively) solitary, while others are meek and clingy. Nevertheless, they manage to cooperate to such a degree that they sometimes appear to be moving as a single body.

The naturalist Mary Austin—who spent almost two decades observing and talking with shepherds in California—wrote that flocks are invariably made up of "Leaders, Middlers, and Tailers." The leaders head up the flock; the middlers keep to the middle; and the tailers chase up the rear. Individual sheep tend to stick to a single role, she wrote, and because leaders can be used to steer the flock, shepherds typically took special care of them, saving them from slaughter to "make wise" the next generation. Some even went so far as to name them after their girlfriends.*

However, in my experience, the flock dynamic was not so simple as Austin describes. There were, rather, many leaders in a single flock, who would arise in different situations. Even more curiously, I began to notice that certain individuals seemed to feel the need to be *perceived* as leading the flock—when the flock abandoned their leadership and changed directions, they would hurry to its front, like a politician scrambling to keep ahead of a shifting electorate.

The relationship between a shepherd and a flock, similarly, is not as clear-cut as it looks. The shepherd is not the master of the flock; instead, the flock and the shepherd are engaged in a continuous negotiation, in turns pushing against each other and pulling together, harmonious one moment and fractious the next. Some shepherds claim to be able to control their sheep with words or whistles, which may be true, but the only signaling mechanism my sheep and I needed was the language of space: if I moved too close to them, they would inch away. In this way, I was able to shape their movements, but only vaguely, like a cloud of smoke. The essence of herding is not domination, but dance.

+

* I have read that Inuit hunters do something similar; they routinely spare the leaders of caribou herds so as to avoid disrupting their migration routes the following year.

Shepherding, like any craft, is a skill acquired over a lifetime—or, ideally, passed down over many lifetimes. My time as a shepherd, by contrast, was a mere stint, a neophyte's first foray. For three weeks in the airy lull of a late Arizona spring, I was stationed near the border of the Navajo and Hopi reservations, in a place called Black Mesa. It was a three-hour drive northeast of Flagstaff, along miles of rutted dirt roads. The area was wholly cut off from municipal electricity, running water, and phone lines. In exchange for herding the sheep, I was given one meal each day and a hut in which to sleep. I had learned about the opportunity from my friend Jake, who had in turn learned about it from an outfit called Black Mesa Indigenous Support, a volunteer organization that helps aging Navajo families remain living on their traditional lands. Jake, who had been shepherding for the past nine years, had regaled me with stories of life among the Navajo, who were among the only people left in North America still herding sheep in the old style, on foot.

A shepherd's life, I learned, is both repetitive and chaotic; like a water wheel, there are whorls within each slow turn. In the morning, just after sunrise, I released the sheep from their corral and worriedly chased them across the hills; in the afternoon, I followed as they galloped to the water trough; and in the early evening, I bullied them back into their corral. At night I slept on a mattress on the dirt floor of a low, octagonal, dome-roofed hut called a hogan. It was part of a homestead that included two hogans, two old stone houses, two new prefabricated trailers, two outhouses, a horse corral, a sheep corral, and the skeletal remains of other hogans long abandoned. There was no running water and no electricity, save a few rooftop solar panels in the main house, which did not appear to get much use.

The whole of it belonged to an elderly Navajo couple named Harry and Bessie Begay. They were both in their late seventies. Harry was a gray-haired man with a talon nose, thumb-punched cheeks, skeptical eyes, and perfect posture, who wore a baseball cap when he rode his horse and a cowboy hat when he went to town. He was missing two fingers on his right hand—lost, I would learn, to the slip of a chainsaw. Bessie, his wife, was a sweet, tough woman who stood no more than five feet tall. She wore velveteen blouses clasped at the neck with a turquoise-and-silver brooch and a black scarf knotted around her tight bun of steely hair. Her mouth rested in a soft frown, except when she found something amusing, and then it lifted to form a smile the exact size and shape of an upturned cashew.

Harry and Bessie may well be the last generation of sheepherders in their bloodline; none of their six living children had plans to return to their ancestral land and scrape out a living raising sheep. The steady decline of shepherding is a source of great concern for many Navajo people, since the practice has long been integral to their cultural identity. Archaeological and documentary evidence suggests that Navajos first acquired sheep around 1598, when the conquistador Don Juan de Oñate brought roughly three thousand Churra sheep to the American Southwest. However, the Navajo oral tradition maintains that shepherding stretches back much further, to the dawn of their existence as a people. "With our sheep we were created," proclaimed a local *hataałii*, or ceremonial singer, named Mr. Yellow Water. According to one particularly vivid version of the Navajo creation story, when the celestial being known as Changing Woman gave birth to sheep and goats, her amniotic fluid soaked into the earth, and from it sprouted the plants that sheep now eat. Next, she created human beings—Diné, as the Navajo call themselves[*]—and sent them to live within the four sacred mountains that still demarcate Navajo country. As a parting gift, she gave them sheep.

For centuries, that gift has shaped Navajo culture, just as water

---

[*] Like many indigenous tribal names, Diné means simply "the people." "Navajo" is a Spanish bastardization of the Tewa Pueblo word *navahu'u*, which means "farm fields in the valley."

sculpts a canyon. Navajos' internal clocks were set to the daily schedule of herding, and their calendars were structured by the seasonal
migration. The introduction of wool radically altered their material
culture, by providing the means to weave lightweight clothing, warm
blankets, and intricate rugs. Their architecture was fortified by the
need to protect sheep from raiders. Pastoralism altered their diet,
their relationship to the landscape, and perhaps even their metaphysics. One Navajo woman told the author Christopher Phillips
that herding sheep informed her understanding of the sacred Navajo
principle of *hozho*, or harmony. "The sheep care for us, provide for
us, and we do the same for them. This contributes to *hozho*. Before I
tend my sheep each day, I pray to the Holy People, and give thanks
to them for the sheep and how they help make my life more harmonious." When a baby is born, Navajo parents often bury its umbilical
cord in their sheep corral, in order to symbolically tie the child to
the sheep and to the land. Indeed, as the anthropologist Ruth Murray
Underhill suggested, in some sense the Navajo people as we know
them—or more importantly, as they know themselves—arrived in
this world alongside sheep.

+

On my first morning of shepherding, I sat on a metal folding chair
in front of my hogan, waiting for someone to tell me what to do.
This was my first mistake: as a rule, older Navajos do not relish the
opportunity to explain things to naive, inquisitive white people. They
would typically prefer the pupil learn through silent observation.
Moreover, Harry and Bessie only spoke Diné Bizaad, the traditional
language of the Navajo people. Their English was extremely limited,
as was my grasp of Diné Bizaad. Unless one of her children was visiting, the only person who could translate for us was Bessie's brother,
a rascally character whose name was either Johnny, Kee, Keith, or
all three. (Navajos are known to accumulate multiple names over

their lifetimes.) When he was around, J/K/K acted as the translator between me and the Begays, but he had left that morning in a pickup truck with his friend Norman, saying he wouldn't be back for five days. I was on my own, the only English speaker for miles.

The hogan, like all hogans, was built facing the east, and the risen sun was on my face. Hearing bells, I turned to see a storm cloud of sheep pouring out of the corral. Bessie walked behind them, leaning on an old broomstick. I jogged over to her. With her stick, she drew a circle in the dust, and then bisected it with a straight line: ϕ. At the top of the circle she drew another, smaller circle.

"Tó," she said, using one of the only Diné Bizaad words I knew: "Water."

Using gestures and a few scattered English words, she made it clear that she wanted me to take the sheep to a nearby windmill, which pumped water from the ground into a trough, let them drink, graze them in a big circle, and then bring them home by nightfall. I had seen such a windmill on the drive in, and, while I didn't know how to get back to it, I trusted that the sheep did. (This was my second mistake.)

The sheep were already streaming loosely across the yard toward the shallow canyons to the northwest, so I ran to my hogan, threw some supplies into my backpack, and jogged after them.

I found the sheep in the weeds just beyond the Begays' yard. They went snuffling along the ground, plucking out tender green shoots of grass, their lips fluttering rapidly. Occasionally, I glimpsed the bright flash of a wildflower before it vanished.

I noticed that this flock had recently been shorn; the khaki folds and fissures of their backs resembled an aerial view of the desert. The Navajo-Churro sheep, the oldest breed in North America, is known for its long, straight wool, which is much prized by Navajo weavers. The breed—which has declined greatly over the decades, due in part to the meddling of federal officials who ignorantly judged them "scrubby," "inbred," and "degenerate"—is also known as the American Four-Horn, because some rams grow four full horns. Though I had hoped to see one, this particular flock contained only castrated rams and, therefore, no such marvelous oddities.

Orbiting the flock were five shaggy mutts. Four of them stuck close to the sheep. The fifth, a brown-furred, sweet-eyed little adventurer, quickly attached herself to me. She stayed on my heels from morning to night; when we sat down for a break, she would rest her chin on my knee. Harry had mentioned to his children that he was thinking of getting rid of her, because she followed humans instead of the sheep, making her useless as a sheepdog. But this habit endeared her to me, and I snuck her pieces of beef jerky when the other dogs weren't looking.

Dogs have been used to herd sheep and ward off predators since at least the Middle Ages. With proper training, sheepdogs can be trained to follow a system of whistles and hand signals to manage enormous herds. The Begays' dogs, however, were not those dogs. They responded to no commands (save the one that indicated it was mealtime) and obeyed no master. Their role, so far as I could see, was to bark at anything that moved, be it a sprinting jackrabbit, a terrified horse, or a passing pickup truck.

I had been warned that the Begays' sheep had a reputation for being "a difficult flock," but as we left the homesite and dipped down into a series of sandy stream beds, they seemed sane enough. (Admittedly, I had very little frame of reference.) After spending all night penned up, they walked with vigor, only stopping to nibble once every few steps. The lambs leaped into the air in fishy wriggles. From time to time the young males paused to buck heads, then jogged to catch up.

When the flock encountered a trail, they sometimes jostled into single file—"stringing," shepherds call this—and broke into a senseless run, their ears flapping up and down, until one of the leaders became distracted by a tasty piece of forage and broke up the race. The geometry of the flock varied according to its speed: As soon as they slowed down, the sheep would fan out into a triangular shape, with the widest part leading the way. When the forage was particularly good, they would slow to a crawl and form a roughly horizontal line, like protesters marching arm in arm. As soon as they sped back up, they resumed stringing. As I watched the sheep running in single file that morning, I quickly realized how and why sheep trails form: it was a matter of speed.

But over time I came to notice that even when the sheep were walking slowly, they sometimes showed a strange, almost idiotic, fidelity to these trails. They liked to graze along the trail's edge until it intersected with another trail, at which point, if I didn't intervene, some or all of them would absentmindedly turn onto the new trail rather than follow their former trajectory. They were apparently happy to follow any trail, anywhere.

According to the rancher William Herbert Guthrie-Smith, when domestic sheep are brought to a new area, they immediately begin to establish a habitat for themselves by creating trails. He watched this process firsthand after he purchased twenty-four thousand acres of rain-soaked New Zealand wilderness in 1882, which he patiently converted into a sheep ranch. The first action of sheep in a new land, he wrote, was to "map it out, to explore it . . . by lines radiating from established camps." The sheep trails snaked outward, skirting around bogs, cliffs, pitfalls, and "blind oozy creeks." Many sheep were reportedly "swallowed up" by the wet earth in this exploration process. But eventually, the trails that failed to reach adequate foraging grounds faded, while the useful ones improved. The radial pattern that Guthrie-Smith describes is common for sheep; sunken paths (called "hollow ways") have been found radiating out from Bronze Age villages in Mesopotamia.

Reading Guthrie-Smith, I began to formulate a two-part theory as to why the Begays' sheep placed such blind trust in trails. In the absence of a shepherd, paths provide the basic guidance sheep need to find their way to food, water, and shelter. As they do for ants and elephants, trails function as a form of external memory. Just as the notion of building a road that leads nowhere seems absurd to us, it would never occur to sheep that one of their trails might not lead to something desirable. So they follow them, trusting that the destination will be worthwhile. At the same time, sheep trails also carve out new sunny spaces (what ecologists call "edge habitats") where different species of grass take hold; in New Zealand, Guthrie-Smith noted that along sheep trails sprouted "succulent green stuff such as white clover, suckling, cape-weed, and sorrel." I would not be surprised if something similar was happening in Arizona, because the sheep showed a preference for grazing along roads and trails (assuming the forage hadn't already been picked clean by another flock). In this simple fashion, sheep use trails to begin bending the land to their needs.

+

In the calmer moments that first morning, I was able to able to admire the desert. The soil was the mingled color of pencil shavings, in turns a pale yellow, a powdery pink, and a dry black. Out of it grew a stiff yellow grass. I recalled John Muir's description of California's Central Valley in late May: "Dead and dry and crisp, as if every plant had been roasted in an oven." Actual tumbleweeds actually tumbled across my path. Things poked at my ankles as I walked: spiky tufts of grass, tiny bamboo groves of the green ephedra plant called "Mormon tea," ankle-high cacti with spines the color of old toenails. The only shade came from the scattered juniper trees, which writhed against an ageless wind.

Off to the northwest, I spotted a windmill, but it looked as tiny as a tin toy. While I was contemplating whether, and how, to turn the flock around, the sheep—as if hatching a whispered scheme—began to divide into two equal-sized groups. I watched the split slowly forming, but I couldn't move quickly enough to prevent it.

One group drifted downhill, off to the east, while the other nosed up the hill to the west. Placing my faith in the directional sense of the leaders—my biggest mistake yet—I focused my attention instead on the tailers, figuring that they would be less headstrong. I broke into a run and skirted wide around them. Then, shouting curses, I attempted to rush them up the hill. But now their gait—which all day had been brisk and light—was suddenly slow, their hooves leaden. They stopped often, glancing about, as if entering unfamiliar and dangerous territory. Growing increasingly panicked that I would lose half of the Begays' sheep, I left the sluggards where they were and ran up the hill in the direction I'd last seen the other half of the flock.

The land rose to a flat tabletop, runneled with narrow washes and forested with pinyon pine. I imagined that sheep were lurking behind every stand of trees, and I even heard the spectral gonging of their bells, but they were nowhere to be seen.

As I reached the top of the mesa, something trotted across my path. It moved from my right to my left, low and quick. For a moment I thought it was one of the dogs.

Then I recognized it: a coyote. Ears up, mouth open, it glided over the sand with the cool certainty of a missile.

A sick feeling bloomed in my abdomen. I envisioned finding one of the lambs torn open, its red chest toothed with white ribs.

Running in a circle, I shouted for the dogs, whose names I did not know. Then I ran back down the hill, where I'd left the other half of the flock, only to find that they, too, had disappeared. It seemed impossible, an elaborate practical joke. I turned in circles, feeling dazed. In my mouth had grown a cat's dry tongue.

The word *panic*, fittingly enough, refers back to Pan, the mischievous goat-legged god whose bellowing used to terrify shepherds and their flocks. Suddenly I felt its true meaning—a blinding electricity that floods the mind, prompting action without premeditation. I ran back up the hill. I found nothing. I ran back down to the valley: more nothing. Then, losing hope but unsure of what else to do, I ran back up the hill.

It was not yet ten in the morning on my first day of herding, and I had lost every last sheep.

+

It is perhaps no accident that the idyllic stereotype of the happy, lazy shepherd—as popularized by poets like Theocritus, Milton, Goethe, Blake, and Leopardi—began to crumble as soon as it reached the wide expanses of the American continent. John Muir, a self-described "poetico-tramp-geologist-bot. and ornith-natural, etc.!!!," spent the summer of 1869 with a sheep outfit in the Sierra mountains as a young man. Most of the time he left the shepherds to mind the sheep while he traipsed around making sketches of glaciers and pines. He hated the sheep (deeming them "hoofed locusts") and had scarcely more respect for shepherds, whom he found to be filthy, intellectually dull, and mentally unstable. "Seeing nobody for weeks or months," he claimed, the sheepherder "finally becomes semi-insane or wholly so." Archer Gilfillan, a herder of nearly twenty years, agreed. "Considering all the things that can and do happen to a herder in the course of his work," he wrote, "the wonder is not that some of them are supposed to go crazy, but that any of them stay sane."

I was beginning to see what he meant.

Looking to the south, I spotted the Begay's blue pickup truck inching along a dirt road. I wondered if they had been quietly tailing me all morning, having anticipated just this sort of debacle. As I approached, Bessie rolled down the passenger-side window. Her eyes were big behind her glasses, her mouth down-curled into a perfect omega. She said something complicated in Diné Bizaad, then, registering my confusion, simply asked: "Where the sheep?" Her voice quavered. I attempted to pantomime what had happened, with poor results. She looked down and fished an old flip phone out of an embroidered pouch around her neck, poked at it a few times, then handed it to me. On the other end of the line was her daughter, Patty.

"Okay, what happened?" Patty asked.

I told the story: fission, drift, the frantic race between two widening poles, then . . .

I handed the phone back to Bessie. Patty translated. With a sigh, Bessie clapped the phone shut and gestured for me to get in the truck.

We slowly prowled the dirt roads. Once every few minutes Harry would stop the truck and they would get out to inspect the ground for fresh tracks. After one of these stops, I hopped up into the bed of the truck to gain a higher vantage (and to avoid Bessie's sightline). I was queasy with guilt. In the matrilineal and matrilocal Navajo society, a family's sheep traditionally belong to the women, and Bessie's deep attachment to her sheep was palpable. They represented not just a sizable chunk of her life savings—ten thousand dollars, more or less—but also decades of labor and centuries of tradition. The sheep I'd lost were a living inheritance from her ancestors, and future gifts to her grandchildren.

After an hour of searching, we gave up and drove home. Patty was waiting there with her two boisterous kids. They were sitting on the couches that lined three walls of the Begays' sunny living room. Patty paused from shushing her children to welcome me back.

"So, how many did you lose?" she asked.

I sighed painfully. "All of them."

"Don't worry, happens all the time," she said. "They show up eventually. Maybe we lose one to a coyote. That happens too. Wouldn't be the first time. Won't be the last."

She had brought a plastic cooler full of raw skirt steak—a welcome treat for Harry and Bessie, since they lived almost an hour from the nearest grocery store and had no refrigerator. She went outside to stoke the wood fire for the grill. I sat in the living room and stared at the walls, which were lined with old family photographs, calendars, and a large tapestry one of the Begay boys had brought back from his stint in the military, which depicted two colonial officers, astride elephants, hunting down a tiger. A bookshelf was stocked with yellow-spined *National Geographic* magazines dating back decades. The couches were neatly covered with bedsheets. Flies circled the room in endless, spirographic patterns.

Some time later, Harry came riding into the yard on his horse, herding half the flock in front of him. Watching them funnel into the corral, I felt some relief, but not much. The other half was still out there with the coyotes.

After lunch, we got back into the truck. Rather than driving west, where I had lost the sheep, we drove due north, on a dirt road that ran up the middle of a grassy valley. At the northern end stood the chrome-bright windmill I had spotted earlier.

Patty pointed out her left window and told me to avoid taking the sheep over there, to the west—the precise place I had taken them. "They get all crazy up in those hills," she said. Plus, she added, it was too easy for a new shepherd to lose sight of them among the trees and gullies. It was better to walk them in wide circles around the valley, a place I would later, in my endless perambulations of its grassy slopes, come to call "the salad bowl." (I recalled Bessie's map, drawn in the dust, which suddenly made perfect sense. It was the salad bowl bisected by the road: ϕ.)

We pulled up to the windmill, which revolved slowly, its innards pistoning, drawing water from the ground. On its tail vane, in red paint, was printed: THE AERMOTOR CO/SAN ANGELO, TX/USA. Beside it stood a trough and a ten-foot-tall holding tank of water.

In its shade stood the sheep.

We counted them: They were all there. None had been eaten by the coyote. The dogs were all nearby. Something in my gut slowly unclenched, and I could breathe. (*Perhaps those dogs aren't so useless after all*, I thought.)

Patty told me to walk the sheep home. "*Slowly,*" she added.

I stepped out of the truck and walked around the sheep in a wide circle. They looked at me placidly, without a wrinkle of guilt. Even a dog would have had the courtesy to avert its eyes, but they were blameless as lumps of snow.

When they had finished drinking, we started off home. The sheep seemed to know the way, so I slung my walking stick over my shoulder and ambled behind them as they crossed the sun-washed valley. Once again, the shepherd's life seemed idyllic.

When we arrived back at the Begays' property, I steered the sheep toward the corral, a shoulder-high enclosure made up of wooden boards, scrap metal, and mismatched plastic tarps. The other half of the flock was already waiting inside the corral, and upon hearing our approach they began bleating frantically. My half of the flock shouted in idiotic response. The moment I opened the corral door to let my sheep in, chaos broke loose: the trapped sheep attempted to escape as the other half attempted to invade. A white liquid roil ensued. Hungry lambs rushed out and swiveled their snouts between their mothers' hind legs, latching onto their udders even as the ewes strode forward into the corral. Despite my best efforts, two hungry sheep escaped. I assumed they would obey their flocking instinct and follow the rest back into the corral, but instead, to my horror, the majority of the flock turned and rushed to follow the escapees before I could shut the unwieldy gate.

The glitch, I realized, was that my half of the flock had already stuffed themselves with grass, whereas the other half, which Harry had rounded up earlier, had spent the afternoon growing hungry.

The escapees slunk off, looking for grass. No matter what I tried, I could not coax them back; I could herd them as close as the corral, but as soon as I opened the gate, the leaders would rear their heads and gallop back out to pasture, trailing the rest behind them. Eventually the two rebellious sheep allowed me to herd them into the corral, but only once they had eaten their fill. This was my final lesson that day. In the words of Muir: "Sheep, like people, are ungovernable when hungry."

+

Over the course of many years, shepherds and their flocks mold to each other. They tailor each other's behavior and shape each other's bodies—the shepherd tries to keep the sheep fat, while the sheep endeavor to keep the shepherd thin. With time, humans weed out the sheep that refuse to follow (by butchering them), and the sheep weed out the humans who are unfit to lead (by driving them to a state of either insanity or depression).

One morning, Bessie went off to run an errand, so she sent Harry off with the sheep and tasked me with preparing lunch. Leaving a pot of beans to simmer, I snuck out to observe Harry at work. I was struck to find that, under his care, the sheep were utterly calm, stopping for long periods of time to pick over the grass, whereas with me they had been as restless as fleas. Being too old and stiff to walk long distances, Harry sat tall atop his horse, a brown stallion with a white star on his forehead. He paced around the flock in graceful curves, slowing down the leaders, hurrying up the stragglers, gently molding the cloud. I never even saw him trot; the horse took slow, balletic steps. When a straggler failed (or refused) to catch up, Harry sometimes circled back for it. Other times, he appeared to leave it behind, confident it would eventually return to the fold.

Over the centuries shepherds have developed many clever ways of managing their flocks. In many countries, shepherds train a goat or a castrated ram, called a "wether," to follow spoken commands. (To more easily locate it, shepherds tend to put a bell on this sheep, a practice that furnished us with the term *bellwether*.) The custom of bellwethering was noted as far back as Aristotle's *The History of Animals*. In 1873, the British writer and magazine editor Thomas Bywater Smithies relayed an anecdote that, under other circumstances, would haunt the nightmares of anyone who has herded sheep: One day, he watched as thousands of sheep from many different flocks mixed together by the banks of the Jordan River. "It seemed a scene of inextricable confusion," Smithies wrote. "But as each shepherd gave his own peculiar call, the sheep belonging to him, and knowing his voice, came out from the crowd, and followed their own leader."

In my three weeks of herding, I did not have time to train my sheep. I was instead relegated to the role of benevolent predator, chasing them to where I thought they should go. As the weeks passed, though, I did pick up a few tricks. I learned not to micromanage the sheep, because (as Moroni Smith, a Utah sheep rancher, once wrote) "an anxious herder makes a lean flock." I learned to see the differences between each member of my flock; I gave each sheep a nickname and started to recognize their individual personalities, which allowed me to predict their movements. I learned that the tailers hung back for a reason—by slowly hunting up the dregs the hard-charging leaders left behind, they filled a niche. I learned that stray sheep are at the greatest risk of wandering off when they are in a large group and feel insulated from danger.* And I learned the importance of setting the sheep off in the correct direction as they left the corral in the morning, since the trajectory of their first hundred steps tended to dictate the following thousand, a phenomenon social scientists call "path dependence."

I also learned why certain sheep stray. Some of my sheep—

---

* The rule of thumb I devised was that any group of seven or more sheep were capable of forming a quorum and wandering off. However, this rule-of-seven was far from ironclad; I once lost a group of five for an entire afternoon.

most notably the one I called Burr Face, a gaunt, knock-kneed old ewe with a large burr permanently fastened to the wool on her left cheek—would routinely wander off. Initially, this just seemed like an error to me, but I came to realize that straying is a calculated gamble. The goal of every sheep is to spend as much time eating and as little time walking as possible (while, in turn, keeping themselves from being eaten). Most of the time, straying was an ill-advised decision, because I would chase the strays back to the herd, which meant they spent more time walking and less time eating. However, in the desert not all foods are considered equal. Grass is a staple for sheep, but what they prefer is pygmy sagebrush, wildflowers, or, especially, the fruit of the narrow-leaf yucca plant. (The very sight of yucca could send even the most indolent of sheep into a mad dash.) Every few escape attempts, the stragglers chanced upon one of these calorie-rich foods. On one occasion, Burr Face staged a small insurrection, leading six other sheep away from the herd toward a large patch of sagebrush. Seeing the wisdom of her discovery, I turned the herd around and led them all back to where she stood—and so, for thirty glorious minutes, a lifelong straggler was transformed into a far-sighted leader.

Most important, I learned that whenever possible, a shepherd should attempt to bend the will of the sheep, rather than break it. By locating the nodes of desire that sheep naturally gravitate toward, I found I could steer the flock without unduly stressing it. Smith wrote that the object of skillful herding is not to bully the sheep, but rather to "create a desire with the sheep to do the things that the herder wants them to do," which, he added, "is the secret of successful handling of all animals."

+

When I was younger I used to see the earth as a fundamentally stable and serene place, possessed of a delicate, nearly divine balance, which humans had somehow managed to upset. But as I studied trails more closely, this fantasy gradually evaporated. I now see the earth as the collaborative artwork of trillions of sculptors, large and small. Sheep, humans, elephants, ants: each of us alters the world in our passage. When we build hives or nests, mud huts or concrete towers, we re-sculpt the contours of the planet. When we eat, we convert living matter into waste. And when we walk, we create trails. The question we must ask ourselves is not *whether* we should shape the earth, but how.

When you herd sheep, those living lawn mowers, this question becomes all the more urgent. A skillful herder with a willing flock can radically transform the ground they walk on, for better or worse. In *Tutira*, Guthrie-Smith describes how, over the course of forty years, he and his sheep converted a tract of land covered in bracken, bush, and flax into a bucolic, grassy sheep ranch. First, the sheep trampled trails through the bracken and manuka, which created canals to drain the bogs and allowed palatable native grasses, like weeping rice, to sprout along the trails' edges. Areas of spongy, fern-choked turf were soon compressed into a soil fit for grass. The sheep's manure fertilized the ground and re-sodded hills blown bare by the wind. The sheep even constructed "viaducts" between hilltops and "sleeping-shelves" on the hillsides. Year by year, they quite literally carved out a place for themselves to live.

However, Guthrie-Smith warned that, if the shepherd isn't careful, sheep can have the opposite effect. When they walk across a plot of land too often, their hooves can compact the soil to an "iron surface," which hinders grass growth. More destructive still is the problem of overgrazing, which is described in alarming detail in Elinor G. K. Melville's *A Plague of Sheep*. When allowed to breed unchecked, sheep sometimes enter a pattern of what ecologists call "irruptive oscillation" (boom-bust), which can permanently degrade the landscape. When too many sheep graze in the same area, they eventually begin chomping grass down to its roots. In warm, dry climates, this can eventually lead to what is known as "ovine desert-ification." The process is deviously self-reinforcing: Grass normally serves to both shade the soil and retain rainfall, so when grass is cropped too low, the soil desiccates. Drier soil leads the existing plant species to die off, and new species—those which are better suited to drier climates and, not coincidentally, inedible to sheep—take their place. As this prickly new plant life spreads unchecked, the sheep hunt down the last of the good forage and a vicious circle forms: less forage leads to more cropping down to the roots, which then leads to even less forage. Eventually, the sheep die off in large numbers and the cycle is broken, but not before the soil and vegetation are irreparably changed.

According to Melville, in the sixteenth century the introduction of Spanish sheep—against the strenuous objections of the indig-enous population—into Mexico's Valle del Mezquital converted a number of grasslands and oak forests to arid scrublands thick with thistles, mesquite, and other spiny plants. By the end of the century, Melville wrote, "the 'good grazing lands' of the 1570s had become scrub-covered badlands."

In the 1930s, the Bureau of Indian Affairs believed that this pro-cess was taking place in the Begays' corner of the Navajo reservation. Between 1868 and 1930, the Navajo population grew fourfold. Their growth was fed by a roughly parallel explosion in the population of sheep and goats, which the Navajo herded along looping routes from the summer highlands to the winter lowlands. However, the soil had begun to dry out, and the good forage was giving way to (aptly named) toxic plants like snakeweed, sneezeweed, Russian thistle, and locoweed. Federal officials believed that if drastic reductions were not made in the Navajo livestock population, the bulk of the reservation might effectively degrade into a wasteland.

At the time, the Bureau of Indian Affairs was headed by a well-intentioned but ultimately tragic figure named John Collier. Raised in Atlanta and educated in New York and Paris, Collier romanticized the Navajo nation as "an island of aboriginal culture in the monotonous sea of machine civilization." However, he also viewed their herding practices—based as they were on tradition, spirituality, and firsthand knowledge—as inferior to the burgeoning science of range management. While it was plain to Collier and his colleagues that the Navajo range was over-grazed, many Navajos believed that the poor forage was merely the result of an unusually dry spell of weather. (Indeed, the region was suffering from the same climatic shift that famously converted the prairies of Oklahoma into a dust bowl.) Some Navajo elders held that the drought was brought on by a breakdown of religious tradition, which, ironically, meant that Collier's proposed plan—of slaughtering huge numbers of sheep—could further upset the Holy People and worsen the drought.

The Navajo, who had lived on the land for centuries before Collier even arrived, were understandably upset by the notion of a white stranger from Georgia telling them how to manage their sheep. Some of Collier's advisors, like the forester Bob Marshall, stressed to him the importance of crafting an approach that respected the Navajos' metaphysical beliefs, complex family dynamics, and profound knowledge of the land.

Collier did not heed this advice. Instead, he instituted a draconian system of stock reduction in which thousands of sheep, goats, and horses were shot en masse. Many corpses were either left to rot or doused in kerosene and burned. In the end, the total number of livestock was cut in half. Moreover, Collier sought to "modernize" the Navajo herding system by breaking the land into eighteen grazing "districts," which disrupted the annual migrations that had long allowed the Navajo to adapt to a harsh and volatile climate. The Navajo Tribal Council frantically passed a number of resolutions in an effort to halt the regulations, but Collier exercised his congressionally mandated veto power. In frustration, many Navajos resisted violently, while others protested to Congress. Finally, by 1945, Collier was ousted, and his livestock reduction plan quietly scrapped. His bitter tenure is still remembered by many Navajos as a period of cultural genocide.

In hindsight, it seems clear that the core problem of the 1930s was not that the Navajo had too many sheep, but that the land had too many people. As the historian Richard White has noted, as the Navajo population continued to grow, an exploding population of Anglo and Chicano ranchers were edging into Navajo-owned lands. Indeed, over the same time span that the Navajo population quadrupled, the total population in Arizona multiplied by a factor of sixty-seven. Throughout the 1930s, government officials repeatedly warned the Navajo that they were facing a Malthusian disaster. However, White noted, no one ever told the Anglos to slow *their* population growth, or cease encroaching on Navajo land. Collier and other federal officials, being unwilling to grant more grazing land to the Navajos, opted instead to cull the Navajo herds and disrupt their traditions. In the process of trying to preserve their land, Collier ended up embodying many of the worst aspects of imperialism—ignorance, racial supremacy, and brutality. Despite his efforts, or perhaps in part because of them, the rangeland's vegetation has continued to wither ever since.

In effect, Collier believed his job was to shepherd the shepherds: for his plan to work, he needed to convince a population of intelligent, independent-minded people to alter their traditions and sacrifice much of their wealth. It was a delicate task that would surely have been better handled by the Navajo themselves. Perhaps, as some have suggested, a Navajo leader could have forged a collective agreement around their core belief in *hozho* (harmony). What's more, any Navajo who had grown up herding sheep would also have understood the most basic axiom of shepherding: though a wise shepherd can bend the flock's trajectory, the shepherd must ultimately conform to the needs of the flock, not the other way around.

+

Two weeks passed. May became June. The sky increasingly took on the blue hue of a butane torch. As the heat intensified, the sheep grew lazier, and the herding became easier. Around two P.M., the flock would gather in the shade of a large juniper tree for a siesta, panting rapidly through flared nostrils. The lambs—who had not yet been shorn—would occasionally become so oppressed by the heat that they would fall forward like drunks and eat while resting on their elbows. Here and there, I would find white balls of wool nested in the grass. When I startled them—and only then—they would sprout legs, spring up, and trot on. By three P.M., every last sheep was heat-stunned, and I would have to chase them from shade tree to shade tree all the way home.

One morning near the end of my stay on the Begays' property, just as I was beginning to gain some confidence in my abilities as a shepherd, Harry and his daughter Jane drove up in a pickup truck carrying five white Angora goats.

After they had unloaded the goats into the corral, I walked over to take a closer look. I peered over the fence and was surprised to find only sheep. Then something uncanny caught my eye. Off to the side stood five strange beings, camouflaged among the sheep, but whiter, shinier. They had tilted eyes and thin limbs, and from their chins hung long white beards. Their nervousness was palpable. I imagined that, to them, the corral must have resembled a prison yard in a foreign land. The sheep's blunt faces and muscled shoulders no doubt seemed brutish, whereas to the sheep, the goats must have looked, as they did to me, otherworldly and effete.

I later learned that this type of goat, the Angora, is highly prized by many Navajo families. Originally from Tibet, the breed made its way to America by way of Turkey, passing through Ankara, where it picked up its name. It is an ancient breed, mentioned in the book of Exodus, but has only been raised by the Navajo in any significant numbers since the turn of the twentieth century, when the goats' long silken hair, called mohair, began to fetch higher prices than sheep's wool. Today, Angora mohair from Navajo country is considered some of the best in the world.

The following morning, when Bessie let the goats out of the corral, I was apprehensive; Jane had told me that the family had tried raising goats before, but had quit because they were too much trouble to herd. As the gate swung open, the first few seconds passed normally. The goats fell in line with the sheep and filed out of the corral. Then the dogs, after an interval of suspicious sniffing, recognized that there were alien beings in their midst, and began barking viciously at the Angoras. The goats flew into a state of panic, skittering away from the dogs with wild eyes. Bessie and I shouted and swung our walking sticks ineffectually at the dogs, who paused from their righteous chase to look up at us with confused and hurt expressions.

I was told by numerous people that goats usually walk ahead of the sheep, but these had the tendency to walk in the back—at times falling so far behind that I would have to circle around and hurry them along. Their hesitation seemed largely to be due to their (understandable) fear of the dogs, who, throughout the day, would periodically forget that morning's lesson, sniff out the presence of these weird not-quite-sheep, and excitedly renew the attack.

The skittishness of the goats threw off my rhythm. It was as if I had spent weeks learning to juggle three rubber balls, and then someone tossed a golf ball into the mix. As we passed through the canyon on our way to the grassy valley, they lingered in areas the sheep trotted past, rearing up on their hind legs to gnaw on flowering cliffrose bushes and low trees. The subtle differences in their behavior made me realize how much I had grown to rely on my ability to intuit the sheep's intentions.

This slight disconnect would lead to calamity. The following day, when the flock reached the far side of the valley, the sheep, as they always did, recognized the windmill, fell into a trail, and galloped for it. But the goats—either not knowing what the image of the windmill signified, or not smelling the water—balked. I decided to follow the sheep, which, if unchaperoned, tended to wander off onto the neighbor's property. (That land was patrolled by a young Navajo man in a black pickup truck, who had angrily scolded me, on two separate occasions, for encroaching on his family's grazing area.) The goats, I reckoned, would either follow behind us, or they would remain where they were.

When the goats did not show up at the trough, I jogged up a hill and caught a glimpse of them in the distance, their wispy tails raised, burning white in the morning sun. Then I ran back, gathered the sheep, and circled around to where the goats had been—only to find that they had vanished. The sun grew hot, pressing the lambs to their knees. Many of the sheep gathered in the shade. I left them there and crisscrossed the valley searching for the lost goats. I looked for hours. Sick once again with shame, I brought the sheep back home and informed Bessie and Harry that the goats had disappeared.

"Oh," Bessie said. We all climbed into the truck.

And so my time herding ended as it began: standing in the bed of a pickup, straining my eyes against the sere hillsides, seeing phantoms in every clump of yellow grass or gap in the trees. From time to time Harry got out of the truck and peered at faint signs printed in the dust, trying to track down what I had lost.

PART III

Hunting

Days later, back in New York, I called to check on the whereabouts of the goats. I was relieved to learn that Harry had eventually tracked down all five, and they were unharmed.

This skill of Harry's amazed me. I had often tried to track down lost sheep or goats myself, but I was never successful. In the talc-fine desert soil, which preserved footprints with surprising clarity, it was impossible for me to differentiate between a track that had been made a few hours ago from one that had been made a few days ago; hoofprints ran in every direction like voices chattering over one another. Harry, however, could easily differentiate between the different tracks with a glance. Indeed, oftentimes he would release the sheep from the corral and allow them to roam free for hours while he attended to some other task. Then, in the late afternoon, he would saddle up his horse and patiently track them down.

Information resides in trails, but it is encoded in a language that must be painstakingly learned. Aboriginal Australians, who are considered by many to be the finest trackers in the world, begin teaching their children to track almost from birth. According to Thomas Magarey, who moved to South Australia in the 1850s, Aboriginal mothers taught their babies to track by placing a small lizard in front of the infant; the lizard would scamper off, and then the child would crawl after it, meticulously tracking it to its hiding place. From lizards, the child would rise in proficiency "until beetles, spiders, ants, centipedes, scorpions, and such like fairy trackmakers are followed over the tell-tale ground." For fun, men of the Pintupi tribe would create startlingly accurate reproductions of animal tracks with their knuckles and fingers in the desert sand, writing fluently in an alien script.

Elsewhere, in the Kalahari Desert, young boys of the !Kung tribe are encouraged to set traps for small game in order to learn about animal spoor. In order to trap an animal, one must predict its future, and the first clue to an animal's future movements is to locate its habitual trails. The simplest traps—crude deadfalls, pitfalls, and foot snares— are often placed along animal trails, a technique trappers call a "blind set." Elaborating on this technique, the indigenous Ndorobo tribe of Kenya dig deep pits, which are sometimes lined with spikes, in the middle of elephant trails. It is an inspired innovation; they locate the elephants' paths, read their futures, and then, like Theseus battling the Minotaur, use the beast's chief asset—its immense bulk—against it.

Following animal trails is the most basic form of what the evolutionary biologist Louis Liebenberg calls "simple tracking." Liebenberg has spent years studying the !Kung people's particular form of persistence hunting, which requires highly advanced tracking skills. As they gain expertise, !Kung hunters graduate to a more "refined" technique, called "systematic" tracking, where a pattern is found and followed among less distinct or discontinuous tracks. Finally, the most complex form of tracking, which Liebenberg calls "speculative tracking," requires the tracker to piece together scanty and scattered evidence to create a hypothesis of where the animal might be headed, so he can expect where to find the next set of tracks.

In his 1990 book *The Art of Tracking: The Origin of Science*, Liebenberg argues that a close study of tracking techniques could resolve a seeming paradox of evolutionary history: How did the human brain evolve the ability to think scientifically—which, in turn, led to an explosion of technology and knowledge—if scientific reasoning was not required for hunter-gatherer subsistence? Plainly, humans did not evolve with the "aim" of one day diagramming the structure of an atom; evolution, as the saying goes, doesn't plan ahead. But then why would humans have evolved the abilities necessary to practice science if we didn't need them to survive?

Liebenberg's answer is simple: tracking *is* science. "The art of tracking," he argues, "is a science that requires fundamentally the same intellectual abilities as modern physics and mathematics." The famed astrophysicist Carl Sagan, who often wrote about the !Kung, agreed. "Scientific thinking almost certainly has been with us from the beginning," he once wrote. "The development of tracking skills delivers a powerful evolutionary selective advantage. Those groups unable to figure it out get less protein and leave fewer offspring. Those with a scientific bent, those able to patiently observe, those with a penchant for figuring out acquiring more food, especially more protein, and live in more varied habitats; they and their hereditary lines prosper."

This theory is an offshoot of an older—and hotly contested—theory in paleontology called the hunting hypothesis, which holds that the pursuit of big game led to much of the development of human language, culture, and technology. I have my doubts about both theories. Liebenberg in particular goes a bit too far in equating "science"—a specific, standardized system of inquiry—with the advanced analytical skills and imagination (what he calls "hypothetico-deductive reasoning") that would eventually allow humans to develop that codified system. Tracking was hardly the only facet of prehistoric life that would have required this skill-set; if tracking is a prehistoric form of physics, then gathering plants is also an early form of botany, and cooking is a precursor to chemistry.

Nevertheless, there is a kernel of truth to these theories: Hunting is an indisputably fundamental human tradition, which has shaped us in various ways. Long before we ever looked at other animals as pets or test subjects, we viewed them as either predators or prey. To understand the full role that trails play on Earth—to see how they can lead not just to a long life, but also to a quick death—I needed to see them through the eyes of a hunter.

+

Until I began researching it, the sum total of my lifelong hunting experience was limited to a few mornings spent in a deer blind on my grandfather's ranch as a child. (My memories consist of staring numbly into the cellarine dawn of a Texas autumn for hours—an exquisitely subtle kind of torture for a child.) So I began asking around in search of an expert hunter. An acquaintance pointed me to a man in Alabama named Rickey Butch Walker. When I emailed Walker, he replied with a concise list of his credentials: with his bow and arrows, he had shot 114 white-tailed deer over his lifetime, and seven that season alone. He didn't use bait, he didn't use dogs, and he had never killed a deer with a gun, in part because he didn't like the noise a gun makes. (He had gotten his fill of guns as a rifle platoon leader in the National Guard, he wrote.) Most important—to me, at least—he only hunted for food, not trophies.

Walker graciously agreed to put me up in his spare bedroom for a weekend and take me out hunting, so I flew down to Huntsville. At the bottom of the escalator to the baggage claim, a big, bullish man was waiting for me. His head shone under the cold cathode lights. It was shaved slick, along with everything else above his collarbone, as if that morning, and every morning, he had placed a razor at the base of his nape and dragged it up over his crown and down the front of his face in one continuous motion. Where his eyebrows should have been were just two wrinkles of muscle. His bright blue eyes were pinched in behind a deep squint—eyes that could be either cheerful or circumspect without changing shape. We shook hands. He took one of my bags. The back of his T-shirt sported the slogan of a brand of hunting clothing called Mossy Oak: "It's not a Passion. It's an Obsession."

Outside the airport it was dusky and warm. We threw my bags into the bed of a red Ford truck and climbed aboard. Walker turned onto a highway, which soon passed over the wide, slow Tennessee River. His cell phone lit up with a text message from his cousin, which I read aloud for him: *I got 1800 lbs of corn today. 36 sacks. My little trailer pulled it great.*

His cousin was apparently planning to use the corn as deer bait, a practice Walker frowned upon. Walker preferred to give the deer a chance. "I like to do it fair and square," he said. "If he can dodge that arrow at thirty yards, that's his business. But if not, I'm gonna put him in the freezer and eat his ass." All year round, virtually the only meat Walker cooked was wild game. When he had a surplus, as he did most years, he gave it away to his elderly neighbors.

As we neared Moulton, the nearest city to where Walker lived, a coyote streaked brightly through the periphery of our headlights. Along the roadside, Walker pointed out some of the twenty-six historical markers that he had researched, written, and installed. He was at least the seventh generation of his family to live in this area—a legacy that, owing to a considerable quantum of Cherokee and Creek blood in his veins, almost certainly stretched beyond recorded history.

After Andrew Jackson chased most of the Cherokee and Creek people out of Alabama in the early nineteenth century, the state imposed fines for anyone caught trading with members of those tribes, so those who escaped the Removal tended to assimilate. Walker's ancestors, many of whom were mixed-race or full-blooded Cherokee, learned to call themselves "black Irish." They were proud, independent folk who lived off the land, never made much money, and occasionally intermarried. Walker often joked that his family tree was more like "a family wreath," because his great-great grandparents had the same grandparents, a fact that is (at least somewhat) less scandalous than it initially sounds: a brother and a sister married a brother and a sister, and then two of their children (first cousins) married each other. Walker said he was not personally opposed to marrying a first cousin, either, though the opportunity had never presented itself.

At the age of sixty-three, he had already quit his job as the director of the Lawrence County Schools' Indian education program to devote himself full-time to bow hunting and studying local history. He had published fourteen histories, eight of which were in print through a publisher out of Killen, Alabama, called Bluewater Publications. "I can talk your ear off about history," he warned me. "I'm a history *nut*."

His grasp of the local history was indeed profound, even overwhelming. He would often begin enumerating a single piece of information, such as the age of a courthouse, but then become lost amid the rhizomatic tangle of ancestry, language, and geography that makes up the Old South. (One afternoon, during a soliloquy on

Indian trails that had somehow veered into an account of the life of the hometown hero and Nazi-conquering Olympian Jesse Owens, Walker caught me dozing off in the passenger seat.) Never in my life had I met a white American who was as deeply rooted in one piece of land. He seemed to know the history of every brick in town. But if he were to drive seventy-five miles away, he would be on terra nova. "I don't know the history of anything else," he said, "but I know the history of my little area."

For him, the wide four-lane roads were overlaid with the onion-skin of history. Instead of Highway 41, Walker saw the Old Jasper Road, a wagon trail that once ran from Tuscaloosa to Nashville. As we turned onto Byler Road, Walker said: "See, now this was called the Old Buffalo Trail. They say you could ride a horse at a full gallop along a buffalo trace and never worry about tree limbs or anything."

Before they were decimated by the rifle and the railway, buffalo had once swarmed the continent from coast to coast, transforming it as they went. American bison (as they are properly known), rather like elephants, tend to walk in single file and can travel great distances. However, unlike elephants, they also sometimes move in oceanic herds; in 1871, Colonel R. I. Dodge encountered a single herd he estimated was twenty-five miles across and fifty miles long. In their endless search for grass, water, and minerals, buffalo created graded trails down hillsides and riverbanks, which became known as "buffalo landings." Where they stopped to wallow, they dug dusty saucers and shallow ponds. They bashed through canebrakes and smashed down groves of Quaking Aspen. Some of their trails were faint, while others were so deep that their shoulders rubbed against the embankments. (A few of these trails, which are still visible in aerial photographs, are so deep that they have been mistaken by geologists for trenches carved out by glaciers.) The Old Buffalo Trail that Walker had pointed out once connected Moulton up to a massive salt deposit called Bledsoe's Lick, where, in 1769, a hunter named Isaac Bledsoe had stumbled upon thousands of buffalo. Around the area's salt licks, the buffalo created radiating trail networks reminiscent of Parisian avenues.

As buffalo trails often do, this section of the Old Buffalo Trail lay on a dividing ridge between two rivers. Buffalo tended to clear trails along dividing ridges, where the walking was easy. Like elephants, they also found the lowest passes over the mountains. When Daniel Boone blazed the Wilderness Road, he followed the path of the Cherokee and Shawnee Indians, who in turn followed the bison, through the Cumberland Gap. In *Rising From the Plains*, John McPhee recounted learning from the geoscientist David Love how bison had discovered the so-called "gangplank," a geological ramp that provided "the only place in the whole Rocky Mountain front where you can go from the Great Plains to the summit of the mountains without snaking your way up a mountain face or going through a tunnel." The gangplank provided the ideal route for the Union Pacific Railroad, which would link the industrialized east to the Wild West.

The geographer A. B. Hulbert wrote that the buffalo "undoubtedly 'blazed'—with his hoofs on the surface of the earth—the course of many of our roads, canals, and railways." However, this neatly teleological account—from bison trails to roads and railways—drastically underplays the role humans played in creating these networks. In many areas, buffalo trails ran in all directions, providing a wealth of options but little direction—historical reports are filled with accounts of pioneers becoming lost in a "maze" of buffalo paths. In other places, there were no buffalo trails at all. What now seems likely is that many of those travelers who marveled at the buffalo's "wonderful sagacity," like Lewis and Clark, had Indian guides who knew which buffalo (or other game) trails to follow, knew which to ignore, and had already furnished paths for the areas where game had not.

The subsequent decline of the bison is well known. As demand grew for their hides, which were used as fur coats and factory belts, white hunters poured westward on the railways, often shooting the buffalo directly from passing trains. Other times, the buffalo were killed by the trains themselves; the sound of an approaching locomotive sometimes prompted them to rush across the tracks in blind terror.

For the federal government, the destruction of the buffalo held a certain monstrously efficient logic: it removed one nuisance (cutting down on the pesky buffalo, which ate up valuable grass, muddied ponds, and derailed trains), while weakening another (depriving the Plains Indians of their staple food source and forcing them to end their roaming existence). President Ulysses S. Grant wrote in 1873 that he "would not seriously regret the total disappearance of the buffalo from our western prairies," as their extinction might increase native people's "sense of dependence upon the products of the soil and their own labors" (i.e., agriculture and capitalism). Millions of buffalo were killed in the 1870s, and by the 1880s they were already becoming scarce. Their bones piled up like snowdrifts, and then were shipped back east to be turned into fertilizer and fine bone china. More than a century later, their absence is felt primarily as a ghostly presence. They are gone, but their trails remain.

+

It was dark by the time we pulled up to Walker's home—a huge, lovingly crafted two-story house, which he had built largely on his own. "I got seven bathrooms in this house," he proudly announced when we walked in from the garage. "Did all the plumbing myself."

He gave me a quick tour of the house's many rooms, flushing the unused toilets in each bathroom so the pipes wouldn't dry out. He was divorced (five times, in fact), and his kids were all grown up, so he lives there alone. On the second floor, he took out a laminated map of the surrounding area and unrolled it on the floor. The map was incredibly detailed; it spanned eight feet from the left edge to the right, but covered less than twenty-two miles. Each hill and hollow was named: Brushy Mountain, Cedar Mountain, Sugar Camp Hollow. "This was my stomping grounds," he said. "I've hiked every creek and hollow in this whole area. Never have been lost back there, that I know of."

Walker had spent his entire life hunting and gathering in one way or another. He pointed out where he used to dig for ginseng, where he used to fish, where he used to probe for mud turtles. As a young boy he had prowled those woods with his tick hound, Blue John, hunting for possums, rabbits, and raccoons. He had set traps for mink and bobcat. Whatever meat he brought home—be it deer, groundhog, possum, raccoon, muskrat, beaver, skunk, squirrel, rabbit, quail, dove, wood cock, snipe, duck, goose, turkey, turtle, bullfrog, or any kind of fish or small bird—his grandmother would expertly cook it up on her wood-fired stove. Pork and chicken were reserved for special occasions.

When Walker was eight his great-grandfather—who was one-eighth Cherokee—had agreed to make him a traditional Cherokee bow and a set of arrows. In his memoir, *Celtic Indian Boy of Appalachia*, Walker describes the process with exacting detail: First, the old man and the boy set out in search of a straight-grained white oak tree that was about two feet in diameter. When they found it, they cut it down with a cross-cut saw, marked off eight feet, and sawed it again. Walker watched as his great-grandfather used a set of metal wedges and a sledgehammer to split the tree in half lengthwise, and then again into quarters. The brittle heartwood was removed and set aside, and the sapwood was split again to create "bolts." His great-grandfather then used a drawknife and then a pocketknife to carve one of the bolts into a bow. (The other bolts were saved for future projects, like axe handles and walking canes.) The old man sanded the bow smooth, rubbed it with brown beeswax, and attached a bowstring made of hemp string, also rubbed with beeswax. Then he carved a set of arrows from slivers of the white oak, fletched them with turkey feathers, and attached steel arrowheads he had forged by hand.

Walker hunted with that bow all through his adolescence. After hunting, he had a habit of storing his bow beneath the tin roof of his grandpa's barn, so that it wouldn't warp. One day, he came back to find that someone had taken it. He was in his twenties by that point, but he nevertheless cried over the loss.

Back downstairs, in the garage, Walker showed me his current bow collection, which hung from screws in the wall. First, he handed me a long wooden bow, which was unstrung. It was made of pale white oak; the grain ran north-south. "This is nearly exactly a replica of the bow my great-grandpa made me," he said. A friend had made him the replica as a gift after his first bow had disappeared, but by that point Walker had already upgraded.

He told me to flex the bow against my thigh as if to string it, to see how "stout" it was. It was like trying to bend a two-by-four. Walker hung the replica back up on the wall. "Now, let me show you what I'm hunting with," he said.

He opened the door of the truck and pulled out a black plastic case, like that of a musical instrument. Inside, nested in foam padding, was a state-of-the-art compound bow. It was muscular and crooked, like a pair of dragon's wings. Its limbs were made of some fancy composite metal and painted with a camouflage pattern. As opposed to a single bowstring, this one appeared to have three: the string was looped through two pulleys (called "cams") at the top and bottom of the bow. Walker handed it to me and instructed me to draw the bowstring. It drew with great reluctance, until the cams began to turn, and then it eased until, near the end, it pulled as smoothly as warm taffy. I looked through the "peep sight," a metal cheerio connected to the middle of the bowstring, to the neon three-pin sight that Walker used to adjust for distance. Normally, Walker wouldn't have been pulling the string back with his fingers; he used a "trigger release"—a device that looped around his wrist, ran along his palm, and hooked onto the bowstring—to ensure a smoother shot. I held the string at full draw for a few seconds, effortlessly, and felt a quivering potentiality of power. Instead of letting go (which Walker sternly advised me against), I awkwardly tried to ease it back to its resting position; the string yanked with surprising force, and the bow nearly leaped from my hands.

"It's like I tried to tell my son-in-law," Walker said. "You need to make sure you get the best equipment, because killing should be humane and quick. Faulty equipment can cause a bad kill, and you can wound a lot of animals."

Inside the plastic case was a quiver full of carbon fiber arrows with razor-sharp expandable broadheads. Walker tested each of the arrowheads against the side of his boot to make sure the expansion mechanism was working. They opened and closed like little silver birds, gliding and diving, gliding and diving.

One of them wasn't opening properly. Walker inspected it and found it was clotted with dried blood, so he rinsed it in the sink. He wiped it dry, and placed it, gleaming, back inside the case.

"I believe if our ancestors"—meaning the Cherokee—"woulda had those bows, things woulda turned out a little bit different," he said. "I would hate to know that I would have to stand forty yards from me with that bow right there in my hand."

+

The next morning we woke before dawn. Walker cooked us a breakfast of venison sausage, biscuits, gravy, and wild muscadine jam. Out in the garage, he was upset to discover that he had forgotten to dry his camouflage overalls the last time he went hunting, and they had "gone sour." The camouflage balaclavas—which would be pressed to our noses for much of the day—had soured, too. We put our kit on anyway, climbed in the truck, and drove to a piece of private land owned by Walker's friend. We parked the truck and climbed out as quietly as possible, being sure not to slam the doors. Walker led me through the woods, sweeping his flashlight left and right to point out the various deer trails that intersected with the main path. He reached down and picked up a handful of empty acorn caps. "That's a good sign," he whispered.

Walker's system was simple. White-tailed deer love acorns, particularly those of the white oak, so he tried to locate a white oak near trampled vegetation and fresh deer droppings. He would then situate himself in a tree about twenty yards away and twenty feet in the air and just wait. "You can just about bet your bottom dollar that within so many hours there's going to be a deer coming to that tree," he said.

"But how do you find that tree?" I asked.

"You have to *walk your ass off,*" he replied.

We stopped at a tree with a ladder bolted to its trunk, a favorite hunting spot of Walker's. Off to the right, a clear pathway skirted around a shallow swamp; Walker preferred this spot because the swamp forced the deer to converge here. Above us protruded a steel bench just barely large enough to hold two men. Walker had chosen to start us off in a permanent tree stand—as opposed to the portable one he usually used, which was essentially a folding chair with steel teeth—because he was worried I would fall and break my neck fumbling with the contraption in the dark. Once we were both up on the bench, Walker strapped his arrows to the right side of the tree and his bow to the left side. Then he strapped me to the tree, via a nylon shoulder harness. I bristled, slightly, at his lack of confidence in me—the bench had a *railing,* after all. Thirty minutes later, I began to nod off, and the strap snapped taut, halting my forward tumble.

We sat for a few hours. The air warmed. The blue leaves grew teal, then green. Wood ducks creaked like old chairs amid the shushing and sighing of distant traffic. From time to time, he tried to summon his prey, almost shamanically, by clacking a pair of antlers against each other and manipulating a small canister, called a "doe bleat," which produced a sound like a remorseful cat. This technique is widespread across hunting cultures: The Penobscot Indians of Maine, for example, used cones of birch bark to mimic the amorous call of the cow moose, while Ainu hunters in Japan used a device made of wood and fish skin to imitate the cries of lost fawns.

Nothing appeared.

After four hours, Walker began packing up.

"Well, we made a good shot at it, but we didn't see shit," Walker said. "You always spend a lot more time waiting than you do shooting, that's for sure."

On our walk out, Walker pointed out more signs: hoofprints in soft mud; a field of clover; a big brown hole in the ground, the rocks around it crusted with dried salt. Deer should have been flocking to this area, but they weren't. Walker's hunch was that they were napping, because they had been grazing all night under the full moon. "When it's a full moon, deer tend to move in the middle of the night and the middle of the day," he said. "It's just a kind of rhythm they go through."

After lunch, we went on a hike in Bankhead National Forest. We were joined by Walker's friend Charles Borden. Beneath his gray beard, Borden had the jarringly youthful smile of a dentist. Like Walker, he wore big leather boots and a T-shirt tucked into his blue jeans. Unlike Walker, he had a black pistol strapped to his leather belt. Behind him trailed a German shepherd named Jojo. He and Walker both carried stout wooden walking sticks as they stomped through the woods, as much to sweep away spiderwebs as for balance.

The two men walked with their eyes pointed toward the ground, like hens searching for seed. Every time Walker found an acorn, he would call out: "Aikerns!" Intermittently, he would bend down to pick one up, crack the shell between his teeth, and inspect the flesh of the nut. A healthy acorn was white and smooth. (Walker handed me one to eat; it tasted like an astringent macadamia.) Sometimes, though, the nut was "faulty," bored through with wormy black holes. Deer can smell faulty acorns.

When Walker found an area with a particular density of acorns, he would glance up, looking for an ideal tree to climb. The trick, he said, was to find a tree that was downwind of the acorn pile, then to climb above the deer's field of vision, preferably concealed by a neighboring tree's lower branches. The two men moved on, looking down, then looking up.

We followed a deer trail that wended gracefully down through the hilly land. One clever hunting method, Borden pointed out, was to find a particularly thick area, then clear a small meadow and a series of paths leading to it, like the spokes of a wheel. The hunter hides near the hub of the spokes, hijacking the animal's trail-following instinct.

Humans are not alone in our ability to exploit the trails of our prey. Many other predators do too: Bobcats crouch in ambush beside game trails; blind snakes can sniff out the pheromone trails of termites; and tiny predatory mites trace the silk trails of two-spotted spider mites. So-called highwaymen beetles, which can recognize the pheromone trails of ants, lie in wait for ants to march past so they can steal their cargo, while green woodpeckers will lay their long sticky tongues across ant trails and simply wait for their meal to be delivered. For most animals, I had come to learn, the ability to make and follow trails provides an evolutionary advantage, until a predator evolves to wield their own trails against them.

Walker paced in circles around an oak tree, looking for signs. "Aikerns ... Aikerns ..." he said to himself, periodically cracking one of the shells between his teeth. As boys, he and Borden had spent much of their free time walking these woods. Both were worried about the fact that their grandchildren wouldn't have the same upbringing. "When you're living in a rural area, and hunting and hiking and staying in the woods, you develop an intimate familiarity with the environment," Borden said. "It gives you a different perspective, because you see the myriad forms of life, and you are able to relate, because you are a part of that. You are not something separate."

+

In the afternoon, Walker brought me to his new favorite hunting spot, in a different part of Bankhead. It was ideal: the ground was well pawed and showered with acorns. It also bordered a small cleared field, which was good, because deer tend to prefer the edge habitats between fields and forests. Walker guided me to the tree he had picked out for me, a tall oak, and began rigging up a device called a tree stand, which resembled the unholy offspring of a folding chair and a pair of crampons. It consisted of two halves, a seat and a footrest, both of which were secured to the tree with a cable that looped around the tree trunk. Each half bore a set of metal teeth that dug into the bark. I attached my harness to the tree and strapped my feet into the footrest. Following Walker's instructions, I began to inchworm up the trunk, first lifting the footrest, tilting my feet back so the teeth bit into the bark, then lifting the top half. Each time I leaned on my elbows to lift my feet, there was an alarming moment when the footrest unclenched from the bark and my feet were suddenly dangling over ten, fifteen, twenty feet of air.

When I'd reached the right height, I cinched both halves of the tree stand to the trunk, folded down the padded seat, and secured my harness. By the time I had finished, I looked up to find Walker already sitting high up in a tree thirty yards away, balaclava pulled down over his face, utterly serene, like a green ninja.

The day slowly undid itself. The air cooled again. Greens blued. Oak toads cheeped, and a coyote let out a neurotic whine. Acorns hailed down onto the ground below. No deer appeared, though. Hunting, I learned, is primarily a battle against boredom. I stared so long into the woods that I began to hallucinate deer out of logs; every falling acorn sounded like a branch snapping under hoof.

Walker sat patiently, his head raptored forward, peering. When a branch fell, his head swiveled and locked in on the source of the sound, then slowly, silently, turned back to center. After a time, he began pulling off beech leaves and letting them flutter to the ground to test the wind. Then he pulled out his cell phone and began texting people, perhaps to relay the non-news. Finally, he began folding up his tree stand and preparing to descend, so I did the same.

In the following days, we fell into a pattern. Each morning we would rise before dawn and return to this hunting spot, where we'd left the tree stands the night before. Around noon, we'd scout for new hunting sites, with an eye for tall white oaks and a confluence of deer trails. And in the afternoon, we'd climb back up our trees and wait.

After three days, we still hadn't gotten close to shooting a deer. As we were driving back from the forest around noon, I asked Walker what he thought goes on in a deer's head. "Well, I ain't in a damn deer's head, but basically what's in their head is feeding, sleeping, and fucking. Same things in everybody else's head," he said. Just then a buck stepped out into the middle of the forest service road about forty yards ahead. The buck's wide eyes and swiveling ears were tuned to our truck, but he stood frozen. Walker stopped the truck and reached for his door handle, saying that he'd "shoot the shit out of it" in the middle of the road if he got the chance. But then he noticed the buck's antlers. They were two stubby prongs. The buck, he explained, was too young to legally shoot. He eased the truck forward to see if we could snap a good picture, but the buck dashed off. After it was gone, we got out of the truck to inspect its tracks, which were accompanied by those of three other deer. Their trail passed through a narrow opening in a laurel thicket, which Walker called a "pinch point."

"See, we fucked up," Walker said. "Deer are probably moving in the middle of the day." By then, this observation had begun to rankle. Every afternoon, Walker would note that we had missed the deer because they were grazing under the full moon, but then every morning he would drag me out of bed before dawn. "It's hard to condition yourself to change your habits," he admitted. "So I guess we get in a rut just about as bad as the deer do."

+

The following day, we awoke again at dawn. We sat for hours, watching the leaves fall. When the sun was high, Walker let out a soft whistle to get my attention. A buck had appeared on the far side of the plowed field. It was a creamy brown, with a white belly and slender legs. It began picking its way toward us along the edge of the field, nibbling the grass.

When the buck was forty yards away, Walker rose to his feet and reached for his bow. If things had gone according to plan, he would have notched an arrow, attached his trigger release, drawn the string back in one fluid motion, and held it at full draw for a few seconds as he lined up the shot. He might have let out a little sound—*Ert!*—to startle the buck and freeze it in its tracks. Then, with a slow constriction of his back muscles, he would have pulled back gently on the trigger until the string released and the arrow leaped silently from the bow, flying at 350 feet per second. It would have slipped in behind the deer's ribs, the arrowhead expanding to cut a two-inch "blood hole" through the vital organs. The buck would have looked startled, then hurt, and begun to limp off into the woods. Rather than following it, Walker would have sat back down and waited for at least an hour; a wounded deer that feels pursued will walk, sometimes for miles, until it falls dead, whereas an unhurried deer will usually lie down within a hundred yards from where it was shot.

Once Walker traced the blood trail to the deer, he would have pulled apart its front legs and cut a small incision beneath the sternum. Then he would have slipped a finger into the hole to push the stomach aside, being careful not to puncture it. (Deer stomachs tend to bloat quickly. "If you cut it too deep and pierce the stomach," Walker warned me, "it will go *FSHH!* and you get shit and chewed-up food all over you.") Wiggling his finger to open a slot for the knife to run, he would have slipped the knife in and cut all the way down to the tail, opened up the chest, cut loose the esophagus and trachea, sliced down each side of the diaphragm, and rolled the guts out onto the ground, where they'd be left for the buzzards to eat. After having cut a hole through the septum of the buck's nose, he would have threaded a stick through it, like a bull's nose ring, and dragged the body snout-first back to his truck.

From there, Walker would drive the dead buck to a special shop for butchering. Years ago, he used to stay up all night butchering his own deer in his backyard with a fine-toothed saw. ("My girls could tell you horror stories about me hanging deer on their swing set," he said.) He would carve the tenderloin into steaks, hand-grind the shoulder into hamburger patties, barbecue the ribs, and save the spine for stew. But that required an enormous amount of time and work, and Walker ultimately succumbed to the gravitational pull of modern convenience. Now, he took the deer to a special butcher called a processor, who usually threw away things like the spine. ("They do it as quick as they can to make money," he said. "And you *let* them do it like they want to do it so you can *pay* the least amount of money.") Since Walker already had a deer in his freezer, he would have given the meat away to his daughters or his neighbors.

At least, that is what *would* have happened—if all had gone according to plan. What happened instead was that, as the buck approached, Walker lowered his bow. "Too young," he whispered. He raised two fingers above his head to indicate that it was the same buck with the Y-shaped horns from the day before. Catching our scent, the buck stiffened, then, after a calculated pause, changed direction and walked in a wide arc around us. Perched up in his tree, Walker's eyes followed it for twenty minutes, as it slipped in and out of shafts of sunlight, appearing and disappearing, passing through the trees at a halting pace. Small and distant, the buck paused, glowing, one last time, and then was gone.

+

Compared to the hunting techniques of Native Americans in the past, Walker's technique was relatively primitive. To better stalk their prey, the Powhatan of the early seventeenth century elaborately disguised themselves as bucks. John Smith described the process in detail: the hunter stuck his arm through a slit in a deer hide while holding a stuffed deer head, "the hornes, head, eies, eares, and every part as arteficially counterfeited as they can devise," Smith wrote. "Thus shrowding his body in the skinne, by stalking he approacheth the Deare, creeping on the ground from one tree to another" until he was within range of a clear shot.

Larger communities tended to rely more heavily on systematic communal hunts. According to Smith, the Powhatan sometimes hunted deer by surrounding them with wildfires and then forcing them to a central ambush point—a widespread technique. The arrival of the fur trade further encouraged mass killing, as opposed to individual hunting.

According to the anthropologist Gregory A. Waselkov, deer was "the single most important meat resource of post-Pleistocene tribes of the eastern woodlands." Put simply, the Eastern tribes could not afford to give the deer a "sporting chance"—a notion wholly alien to them. Though recreational hunting existed in numerous ancient empires, the sport as we know it was only codified by European royalty in the last thousand years.

Deer meat was also the single most important meat resource for the European aristocracy, but for different reasons. Venison was a marker of status, a sign of manly virility, and an indication of geographic power. Deer were so integral to the notion of hunting that they became, quite literally, synonymous with it. According to the historian Matt Cartmill, the modern Irish verb *fiadhachaim*, "to hunt," literally means "to deer-atize." In English *venison*, which originally meant "meat gotten by hunting," now means "deer flesh."

Hunting was meant to serve as a relief from the tedium of the court, so it was ironic, but not surprising, that a baroque system of courtly manners soon arose around the sport. In Elizabethan England, according to Cartmill, "a public spanking with the flat of a hunting knife" was the customary penalty for breaking one of the sport's many rules, "for example, uttering the forbidden word 'hedgehog' during a deer hunt." British royals hunted on horseback, attended by brush beaters, bow handlers, and buglers. In France, the parforce hunt, in which the quarry was essentially run to death by dogs and horse-bound riders, became the norm. Nevertheless, some kings, like Louis XV, managed to kill lavishly. Over his fifty-year sporting career, he is said to have run down some ten thousand red deer, an accomplishment Cartmill considers "possibly unique in the history of the human species."

The royal hunt created a new type of landscape. In 1066, William the Conqueror invaded England, took the throne by force, and began to radically redistribute the land by means of a process called "afforestation," whereby large tracts of land were declared royal forests. Though prior residents were allowed to continue living on these lands, hunting, trapping, herding, and logging were outlawed. The famous New Forest—an ancient woodland and heath in southern England, which remains largely intact—was the result of William's dictate. At their peak, royal forests accounted for one-third of the land in England.

These protections did not stem from some kind of proto-environmentalist sentiment, though. Rather, it was meant to protect the king's prized prey: the red, roe, and fallow deer. William's "forest law" showed a rather sophisticated understanding that without a stable forest ecosystem, large game like deer could not thrive. However, the system was explicitly designed to maximize deer populations; predators did not receive the same protection. Wolves had a royal bounty placed on their heads, and by the 1200s they had been successfully hunted out of southern Britain.

The regulations proved onerous for local residents. William prohibited bows and arrows within the royal forests, and he ordered that all large dogs living near his forests have three claws from their forefeet removed—a grisly procedure, called "lawing," which was performed with a mallet and chisel—to prevent them from chasing his deer. Poachers faced losing their hands, their eyes, or their lives.

As predation of wild deer increased and their habitat shrank, private parks were built by noblemen to protect their stocks of deer, and harsher regulations were passed. Naturally, the common people chafed against these new restrictions. In 1524, three yeomen snuck into a deer park, hacked two young deer to pieces, and ripped two fawns from their mothers' wombs, leaving the carcasses where they lay—apparently, an act of pure, bloody rage. In the oral and written literature of the time, there arose a certain heroic aura around poachers like Adam Bell, Johnie Cock, and most famously, Robin Hood. The famed bandit of Sherwood Forest (a hundred-thousand-acre royal preserve) and his Merry Men represented both rebellion and idyll. They dined on delicacies like venison pasties and sweetbreads. They escaped capture with a superior knowledge of the land, moving, wraithlike, on "derne" (secret) paths. A bounty—called a "wolf's head," because the reward was the same as for that of a wolf—was placed on Robin Hood's life. The poor championed his fight against (among other things) the excesses of sport hunting, while the nobility derided subsistence hunters like him, now dubbed "poachers," as uncivilized and unmanly.

The British eventually brought these mores to the New World, judging Native Americans for their "savage" methods. On seeing the popular Native hunting technique called "lead hunting"—where hunters would locate the seasonal migration route of ungulates, like caribou, and then wait for their prey to appear—the famed British hunter Frederick Selous wrote that he felt "thoroughly disgusted with the whole business. In the first place, to sit on one spot for hours lying in wait for game, is not hunting, and, although under favourable conditions it may be a deadly way of killing Caribou, it is not a form of sport which would appeal to me under any circumstances."

From these elitist roots—predicated as much on aesthetics as conservation—arose the so-called sportsman's code, which frowned on the killing of female and young deer, discouraged hunting for profit, and banned year-round hunting. During the late nineteenth century, these values were enshrined in law by hunter-conservationists like Theodore Roosevelt and Madison Grant, who were instrumental in the formation of many of the earliest national parks and national forests. At the same time, the American sport hunters, in response to the princely diversions of Europe, fashioned themselves as rough and rustic outdoorsmen.

Walker often told me he hunted with a bow and arrow because he was "a strong traditionalist," a man who enjoyed carrying on the rites of his ancestors. As he sat in that tree—in a national forest, in a region where white-tailed deer had been hunted to a vanishing point and then "reintroduced" on two separate occasions over the last century, once a hillbilly boy of Irish and Indian heritage, now a middle-class man bound both by law and honor, staying his arrow because a buck was too young to kill—he was the inheritor and embodiment of more traditions than perhaps even he knew.

+

After our last morning of hunting, when Walker had spared the young buck, he gave me a ride back to the airport. He seemed a bit chagrined that we hadn't shot anything, but overall he was surprisingly equanimous about it. "You aren't guaranteed anything," he said. "That's the reason it's called hunting, not killing."

On the long drive to the airport, to pass the time, he tried to list all the animals he had observed making trails. The list was extensive. "Of course, deer make trails like crazy," he said. "Hogs, that's another one makes bad trails. When they get in a row, they trot, going from place to place, like a line of ducks. I've had trails of them wore out a foot deep.... Snakes leave a trail when they cross a cotton patch. Especially rattlesnakes. We'd track them down and kill them bastards ..."

This was not the first time I'd heard Walker mention an animal making "bad trails," but it jarred me nevertheless. Some environmental preservationists I've met frown upon human trails, which they view as blemishes on the land, but they tend to regard animal trails as natural and good. As a lifelong hunter, Walker sees things differently. He seems to make no differentiation between the trails of humans or any other animal—a rut is a rut. And as the last three days had shown, for the hunter or for the hunted, a rut can be one's downfall.

He went on to list various other trail-makers: raccoons, skunks, turtles, muskrats, minks, armadillos . . .

"I reckon nearly every kind of animal will follow a trail, because it's just easier to navigate," he remarked. "Like the buffalo trails. Most of the time, it's easier walking."

"But," he added after a lengthy pause, his voice brightening to a new thought, "I guess people leave the most obvious trails. Like this damn interstate highway here. Shit, if people cease to exist, somebody could come back here ten thousand years later and probably find remnants of this concrete bridge. So we leave the most destructive trails, I think, of any group of animals."

CHAPTER 4

ONE FROST-LACED fall morning, I went trail hunting with a historian named Lamar Marshall. He was slowly piecing together a map of all the major footpaths of the ancient Cherokee homeland, and he had a new route he wanted to inspect. Wrapped in layers of warm clothing, which we would gradually peel off as the day wore on, we walked down a gravel road through the forests of the North Carolina foothills. The sky was pale, cool, and distant. Down the hill from us ran Fires Creek, which slid southward to meet the fat, muddy tail of the Hiwassee River.

Few Americans can say with any certainty that they have seen an old Native American trail. But almost everyone has seen the ghost of one and even traveled along it. For example, Marshall told me, the highway we'd taken to get to reach these mountains had once been a noted Cherokee trail, stretching hundreds of miles from present-day Asheville to Georgia. The next road we turned onto had been a trail once, too. As had dozens of other roads in the surrounding hills.

Marshall estimated that eighty-five percent of the total length of the old Native American trails in North Carolina had been paved over. This phenomenon generally holds true across the continent, but more so in the densely forested east. As Seymour Dunbar wrote in *A History of Travel in America*: "Practically the whole present-day system of travel and transportation in America east of the Mississippi River, including many turnpikes, is based upon, or follows, the system of forest paths established by the Indians hundreds of years ago."

That system of paths is arguably the grandest buried cultural artifact in the world. For many indigenous people, trails were not just a means of travel; they were the veins and arteries of culture. For societies relying on oral tradition, the land served as a library of botanical, zoological, geographical, etymological, ethical, genealogical, spiritual, cosmological, and esoteric knowledge. In guiding people through that wondrous archive, trails became a rich cultural creation and a source of knowledge in themselves. Although that system of knowledge has largely been subsumed by empire and entombed in asphalt, threads of it can still be found running through the forest, if one only knows where to look.

+

Marshall did not look like any historian I had ever met. He had leathery skin, gray stubble, and two wide-set, sun-narrowed dashes for eyes. From crown to cuff, he wore mismatched camouflage: a camo trucker cap, a camo backpack, and a camo karate gi over a pair of camo cargo pants. Any time I wanted to hear about a new chapter of his life, I needed only point to a garment and ask if there was a story behind it. His trucker's cap read "Alabama Fur Takers Association," an organization for which he, a former trapper, used to serve as the vice president. Around his neck, he wore a beaver skin pouch he'd bought while stocking the trading post he used to run. Beside the pouch hung a sterling silver medallion, which depicted a flattened musk turtle. The turtle—an endangered species, long sacred to the

Cherokee—was the symbol for an activist organization he founded in 1996 called Wild Alabama. That outfit later expanded into an influential conservation group called Wild South, whose efforts currently cover eight southeastern states.

The karate gi was an item he had designed for himself many years ago. He'd since quit practicing karate, having finally decided that "if some three-hundred-fifty-pound guy was going to beat me to death, I'd rather just shoot him." In his former life as a firebrand environmentalist in Alabama, for self-protection he had taken to carrying two powerful handguns everywhere he went. On our hike, to cut down on weight, he only carried a pocket-sized .22 Magnum. "I feel kinda naked with just this," he said at one point, holding it in his palm.

In an orange waist-pack, Marshall carried a GPS device, a few maps, a black notebook, a pen, and firestarter for emergencies. As we walked, from time to time he pulled out the GPS, consulted his map, and took a few notes in his pocket notebook, which was full of hand-drawn maps. He still wrote in the cribbed, cryptic shorthand he'd learned while working as a plat technician for surveyor crews. On the first page, in a gesture reminiscent of the old explorer's journals, he had written his name, and beneath it, his Cherokee nickname, *Nvnohi Diwatisgi*, which means "the Road Finder." (The word for *path* and *road* is the same in Cherokee: *nvnohi*, "the rocky place," a place where the soil and vegetation have already been worn away.)

"Everything gets mapped, everything gets drawn, all the waypoints, contours," he explained. He flipped through the pages. "Every trip since I've been up here: Little Frog, Big Snowbird, Devil's Den Ridge . . ."

Marshall shuffled between copies of historical maps and hand-written historical accounts. On one large modern map, he showed me the trail we would be hiking that day. It ran beside Fires Creek and up over Carvers Gap, connecting the old Cherokee settlements of Tusquittee Town and Tomatly Town. Our walk was only the iceberg's tip of the trail-finding process: the bulk of the work consisted of archival research. He regularly drove to libraries across the country, including the National Archives in Washington, DC, where he and an assistant would spend days paging through old records and snapping digital photographs by the thousands. Once he had confirmed the location of a trail in the historical record, he would use a digital mapping program to plot a tentative route. Then he would hike through the woods searching for it. If he found a trail on the ground that followed his hypothetical line, it was a good indication that it was the old Cherokee trail, but he would still have to perform a transect, walking in a straight line from ridge to ridge, to see if the area contained other potential candidates. "If there's ten trails in there, you say all right, which one was the real trail?" Marshall said. "But if there's one trail in there, then you're pretty sure that's it."

He also paid close attention to the surrounding area, to discern if it was an untouched Native American path, or whether it had been converted into a wagon road, a fire line, or a logging road. (You can identify wagon roads, for example, because they are wider and deeply rutted; you also tend to find piles of rock lying beside them, marking where the road builders tried to flatten the road surface.) Sometimes, he would find three iterations of a trail—the original trail, a wagon road, and then a modern road—laid out side-by-side, like afterimages.

Though his research was best known for helping reveal the startling degree to which our road network was inherited (or more accurately, purloined) from Native Americans, Marshall's top priority was to find those few remaining ancient Cherokee trails that had remained undisturbed. His motivations were (at least, in part) environmentalist: if he could locate a historical Cherokee footpath, federal legislation mandates that the Forest Service must protect a quarter of a mile of land on either side of the trail until it has undergone a proper archaeological survey (which, in certain cases, can take decades). And if the site is ultimately found to be historically significant, then the state can take steps to ensure that the trail's historical context—which just so happens to be old-growth forest—remains intact. By locating and mapping old Cherokee trails, Marshall had so far been able to protect more than forty-nine thousand acres of public land from logging and mining operations.

Marshall's work shook up certain fundamental assumptions about the nature of conservation work. Conservationists generally fight to protect *blocks* of land, whereas Marshall fought to conserve geographic *lines*. Since Cherokee paths often followed game trails, they provide ideal corridors for wildlife to move between ecosystems. The paths also tend to travel along dividing ridgelines, which provide scenic overlooks for future visitors. Even more radically, by showing that human artifacts can serve as the linchpin of wilderness areas, Marshall was bridging an old divide between culture and environment. That dichotomy is familiar to Americans today, but it would have been wholly foreign to precolonial Native Americans. Mile by mile, Marshall was incorporating the human landscape back into the natural one.

+

We walked down the dirt road, looking for openings in the trees. Soon we discovered a trail branching off to the right. It was lovely—open, airy, carpeted with the duff of the overhanging cherry trees, oaks, and pines—but to Marshall it felt not quite right. For one thing, it was too wide. The Cherokee trained themselves to walk heel-to-toe, like tightrope walkers. As one Cherokee man explained to me, "There's no need for a big wide road. All you're going to do is go there, and the things that are there"—plants, medicine, game animals—"won't be there if you make the road wide."

Marshall ventured a guess that this trail was once widened by loggers, and that it would narrow as we neared the top of the ridge.

Five minutes later, though, we ran across a blue plastic rectangle nailed to a tree—a blaze. Marshall's confusion deepened; the trail wasn't supposed to be designated as a hiking trail. And yet, when he consulted his map and the GPS, it appeared we were on course.

He finally concluded that the Forest Service must have appropriated the Cherokee trail. This was unusual. Native American trails normally don't grow into hiking trails, because their objectives differ. Native trails reach their destinations as quickly as possible, sticking to ridgelines while avoiding peaks and gullies. In contrast, recreational trails, which are a relatively modern European invention, dawdle along, gravitating to sites of maximal scenic beauty—mountaintops, waterfalls, overlooks, and vast bodies of water. Modern hiking trails are also meticulously designed to resist the erosive power of hikers wearing rubber-soled boots; so, for example, on a steep hillside, they will cut long switchbacks to lessen the incline. Native trails almost never do this. They tend to charge up slopes in a straight line, following the "fall line"—the path water would take while flowing downhill.

Though Native trails prized speed over ease (and erosion resistance), they also often detoured from the most mechanistically efficient route, for reasons specific to each culture. Gerald Oetelaar, an archeologist who studied the Plains Indians of Canada, told me he became frustrated whenever colleagues relied on computer programs to map "least cost pathways" across ancient indigenous landscapes, because they were laboring under the false assumption that Native people traveled like the Mars Rover, rolling across an unpeopled and unstoried landscape. "I keep pointing out to them: All landscapes have histories!"

Among all living things on Earth, humans are, as far as we know, uniquely rich in what we call *culture*—art, stories, rites, religion, communal identities, moral wisdom—and our trails have grown to reflect this. "There are reasons why we don't do things the 'logical' way," observed James Snead, an archaeologist who studies "landscapes of movement" in the American Southwest. Another way of framing this point would be to say the logic of human behavior is fantastically multiform, as are the trails it creates. A trail might go to great lengths to avoid enemy territory or detour to visit kinfolk; it might gravitate to sacred sites, or bend around haunted ones. Marshall had located a precolonial trail leading up to the crest of Clingmans Dome, where ceremonies were apparently held. If the Cherokee had been following the path of least resistance from one village to another, they would have avoided the mountaintop altogether. Elsewhere in North America, archaeologists have discovered that Native paths often veer to pass close by ritually significant sites, allowing walkers to stop and pray on their way to their destination.

Sometimes the trails themselves became cultural artifacts, much like pieces of art or religious relics. Out west, many tribes used tools to carve trails into the dry soil or stone, like giant petroglyphs. In Pajarito Mesa in New Mexico, Snead found trails running parallel to one another, redundantly, like the tines of a comb; he theorized that the construction of the trails, distinct from the walking of them, held some special significance. The Tewa people built paths, called "rain roads," from mountaintop shrines down to their villages to direct the rain to their crops. The Numic and Yuman cultures constructed paths leading to certain mountaintops (sites of power, or *puha*), which they believed were traveled not only by the living, but also by the dead, dreaming people, animals, water babies, and the wind. These trails existed both in the physical world and in the world of spirits and stories—two different landscapes that, among many Native American cultures, are inseparably entwined.

+

As the trail began to ascend the ridge, Marshall became more certain that it was an old Cherokee trail and not a modern addition. For one thing, it followed the ridgeline, which is a telltale feature of Cherokee trails. He explained that once a walker was high atop a ridge, it was possible to walk for "miles and miles and miles" without encountering serious obstacles. In wintertime, the ridges saved a walker from having to cross through frigid waters, and in summer, they stayed high above the low-lying thickets of ivy, laurel, and rhododendron, which the locals call "laurel hells."

The trail tilted upward, slowing our progress. Marshall calculated that for every mile we hiked, we climbed a thousand feet. He said, between huffing breaths, that this was another good indication that the trail had been made by Cherokees and not Europeans. The English hated Cherokee trails, because they were too steep to follow on horseback.

Though we often speak of the "path of least resistance," a single landscape contains countless *paths* of least resistance, depending on the mode of transportation. The Plains Indians carted goods using a sled-like device called the dog travois, so their trails gravitated to areas of slick grass, like prairie wool, and avoided steep inclines, because the travois would lift the dog's hind legs off the ground. After Europeans introduced horses to the Americas, some tribes also began using a horse travois, which can climb steeper inclines than the dog travois. However, horses cannot climb as steeply as llamas, which meant that farther south in Peru, Spanish conquistadors could not follow many of the Inca trails.

The Cherokee traveled primarily on foot, wearing soft-soled moccasins that allowed their toes to grasp the ground. "The footwear was intimately connected with Indian trails," Marshall said. "It's an aspect that nobody thinks about." On his feet, he wore a battered pair of rubber-soled hiking shoes, halfway between a boot and a cross-trainer, with seams held together by yellow, foamy glue. He had tried wearing moccasins before, but he discovered that his feet weren't strong enough to grip the ground effectively.

The trail rose higher through the brightening air. Gray trees held the husks of dead leaves, shakily. On the side of the trail lay the remains of a chestnut tree, hollowed out by fungus. Chestnut trees were once the most abundant in the region; each summer, they showered the Appalachians in flurries of pale blossoms, and they grew so large that when they toppled over, the sound was known as "clear day thunder." But around the turn of the twentieth century, they started becoming infected with an invasive blight and died off by the millions.

In this and a hundred other ways, the forest we were walking through would have been unrecognizable to the ancient Cherokee. Tyler Howe, a Cherokee historian, pressed this point home when I spoke with him. "The forests today are nothing compared to the forests then," he said. "The natural environment of the Cherokee world has been completely changed." For one thing, nearly all the land had been intensively logged, so the trees would have looked shockingly young to an ancient Cherokee. Moreover, the Cherokee regularly burned the woods, which would have cleared out many of the thickets of rhododendron and multiflora rose, so, to them, a modern forest would look sloppy, unkempt.

The first European visitors to North America were stunned by the forests they found—not just by the age and grandeur of the trees, but also by the lack of undergrowth. Early observers frequently noted that the forests of the Eastern seaboard resembled that of an English park. Some stated that a man could ride a horse (or according to one source, a four-horse chariot) at full gallop through the trees without a snag. A great many colonists ignorantly assumed that this was the natural, divinely ordained state of the forests. Indeed, it may well have appeared that way, because infectious diseases, imported by the earliest explorers, had already killed off as much as ninety percent of the indigenous population before settlers arrived en masse. Those second-wave pioneers had stepped into a vast garden, it seemed, with no gardener in sight.

Even early on, though, observant Europeans cottoned to the fact that the park-like appearance of the forests was the result of careful maintenance. William Wood, who published the first comprehensive natural history of New England in 1634, noted that "in those places where the Indians inhabit, there is scarce a bush or bramble, or any cumbersome underwood to be seen in the more champion ground."* Meanwhile, he noted, in those places where Native communities had died off from plagues, or where rivers prevented wildfires from spreading, there was "much underwood," so much so that "it is called ragged plain because it tears and rents the cloths of them that pass."

In addition to easing foot travel, fire was used to clear farmland, to hunt, to encourage the growth of berry bushes and deer grass, to drive off mosquitoes, and to deplete the natural resources of neighboring tribes. When the British put an end to the practice of strategic burning, millions of acres of open oak savannas reverted to dense forests within two decades. It is now widely understood that, rather than existing in a blissfully "natural" state, the native inhabitants of North America thoroughly altered the landscape, patiently molding it, as a foot breaks in a new moccasin—and being molding by it, as a moccasin toughens a foot.

+

We stopped for lunch at the top of the ridge, where the trail crossed a dirt road. Off in the distance the mountains were isoprene blue. White sun filtered down through high clouds, as sweet and clear as ice melt.

Marshall opened his backpack and pulled out five different plastic baggies. One had a baked potato in it, wrapped in aluminum foil, still warm. Another held an apple. Another, a peanut butter sandwich.

---

*"Champion," also called "champaign," is an antiquated term for open, level country. It has the same root as "campus."

Another held pale cloves of raw garlic, which Marshall popped into his mouth and crunched without grimacing. Another held a slab of blackened bear meat. He had smoked it for two hours then broiled it in the oven to leach out the remaining fat. He cut me off a piece. It was delicious, reminiscent of Texas smoked brisket. For himself, Marshall saved a huge bear rib, which he gnawed at like a wild, white-muzzled dog.

He lay on his side, propped on an elbow, telling stories from his youth. When he was in fifth grade, he said, he became obsessed with stories about American Indians; he would hide recollections of frontier life inside his textbooks so he could read them while pretending to study. Naturally, he gravitated to the Boy Scouts, where he learned to hike, canoe, and camp out. When he was eighteen, he built a raft out of fifty-five-gallon drums (complete with a sail, a detachable canoe, and a ten-foot Confederate flag), which he and two friends floated down the Alabama River from Selma to the Gulf of Mexico.

Soon after, he befriended an "old mountain man" named John Garvin Sanford. As the two went "prowlin'" through the woods in search of ginseng and goldenseal, Sanford would sometimes lead Marshall to the site of old Cherokee villages. On one occasion, Sanford dug down into a fire pit in an abandoned village and recovered a pile of tiny, charred corncobs. (Ears of corn, he explained, were much smaller before Europeans began cultivating them.) Marshall canoed to various former townsites to see if he could find shards of pottery or remnants of tomahawks. Sometimes, standing in a plowed field, he could see the dark circles and squares where Cherokee houses had once stood; even after being tilled countless times, the ground was still blackened from centuries of cooking fires. He puzzled over the old Cherokee trails, where they went, and why.

In the following years, he drove across the country, hiking and canoeing the wildest places he could find. When his friends went off to Woodstock, he went to Canada to paddle the lakes of the Quetico.

He took up studying survival skills and opened a survival school called the Southeastern School of Outdoor Skills. For two years he made a living trapping mink, muskrat, raccoon, and fox. "Everything the Native Americans did, I wanted to emulate," he explained. "I saw the whole world through them."

For a long time, he had believed he was one-sixteenth Cherokee, but in 2015 he took a DNA test that suggested otherwise. "Family traditions sometimes are found to be family fantasies, I guess," he later wrote to me in an email. More disappointing still, he had discovered that one of his ancestors, while fighting in the Revolutionary War, had helped burn the towns of the Lower Cherokee, who were then allies of the British. "I guess that my mission in life," he concluded, "is to make retribution for the sins of my ancestors."

In 1991 Marshall's passion for wild lands began to take on an activist edge. That year, after many years of working as an engineer for corporations that built paper mills and nuclear power plants—work he despised—he purchased a 140-year-old cabin on an inholding in the Bankhead National Forest. He moved there in the hopes of getting back in touch with the wilderness, but on his regular hikes, he was horrified to find that huge patches of the forest, including stands of old-growth trees, had been razed.

One day, he ran across an article in the local newspaper by none other than Rickey Butch Walker denouncing the clearcuts in Indian Tomb Hollow, a site containing ancient Cherokee steatite pottery. Marshall befriended Walker, who showed him the desecrated site. At first Marshall was enraged by the damage he saw, but then he had an inspiration: He decided to cobble together a newsletter called the *Bankhead Monitor,* which would chronicle the ongoing destruction of the forest. The front-page headline of the first issue read, "Alabama Chainsaw Massacre: Clearcutting a Historic Site." (He surreptitiously printed the first copies of it at work, in the office of Amoco Chemicals.) Marshall began by handing the newsletters out for free in parking lots, then he sold them for a dollar in local stores. They gradually caught on. Over the course of fifteen years, the four-page newsletter grew to a full color, one-hundred-page magazine with a circulation of five thousand.

In 1994 an anonymous donor offered to pay Marshall a yearly salary to ensure that he could quit his job and fight the Forest Service full-time. Marshall accepted the offer, redoubling his efforts. However, selling environmental protection to rural Alabamians proved a tricky task. At one community meeting, in a remote country church, Marshall narrowly avoided a mob beating, thanks only to the intervention of a local preacher. On another occasion, Marshall and two friends were held at gunpoint by an inebriated hunter, who ranted about how environmentalists wanted to "lock up the forest." (They managed to escape only when the hunter bent down to draw a map in the dirt to show the location of a nearby well—where he intended to dump their bodies, Marshall presumed—and toppled over backward.)

As the fight intensified, Marshall began receiving death threats. He took to wearing two guns whenever he was in public, a 9mm Glock and a Smith & Wesson .357 Magnum. His then-wife bought a gun as well. For a time, he hired off-duty police officers to guard his property. Local residents boycotted his business—a small country store called the Warrior Mountains Trading Company—which he was eventually forced to sell. All told, he lost $400,000 over the course of those years, the bulk of his life savings.

Marshall had started out life "as rightwing as they come," he says: In his twenties, he was a member of the John Birch Society and a campaign volunteer for Ronald Reagan. In the years since, his beliefs had occasionally drifted leftward, but not by much. "Conservation," he liked to say, "is conservative." It came as a shock, then, when his opponents in the fight over Bankhead National Forest tried to paint him as a leftist radical. "It was like I was the most evil, liberal, godless person ever to exist," he recalled. "They called me a communist!"

Marshall eventually came to describe himself as a "conservationist" rather than an "environmentalist." Within the larger environmentalist community, he was something of an enigma, he said. As a Christian and a sportsman, the biocentric approach of Deep Ecology, then in ascendance, held no interest for him. He loved the woods because it was a place for humans to roam free, to hunt, and to fish. Environmental activists from the North seemed to come from a different planet. At one Greenpeace training camp he attended in Oregon, he was surprised to realize that he was virtually the only person in attendance who ate meat. While there, he took a workshop on how to climb trees and hang protest banners. A public radio reporter asked him if he was going to use those skills when he returned home to Alabama. "Oh hell no," he replied. "If you climbed a tree in Alabama, they'd cut the tree down. If you chain yourself across a road they'll run over you. You can't do that kind of stuff in Alabama."

In the end, it was Rickey Butch Walker who cracked the code of how to convince Alabamians to fight for their wild places. Having grown up in those woods, Walker knew that for many people, the wilderness did not represent an otherworldly sanctum of 'biodiversity,' as it did for many urban environmentalists. Rather, it served as the birthplace, staging ground, and repository for the area's deepest traditions. Walker urged Marshall to shift his focus from protecting endangered species to protecting local traditions. Hunting and fishing were considered sacrosanct, and, in an area where roughly a quarter of the residents claimed some form of Cherokee ancestry, historic tribal sites were fiercely guarded. Marshall quickly saw the merits of Walker's approach. "You go on down to Alabama, and people don't give a damn about endangered squirrels or whatever," he told me. "But if you go up there and want to mess with that hill where they killed their first deer, boy, they'll *kill* you. Everything has to be framed in personal language. The more educated people are about their roots, the more connected they're going to feel to their land. And then they're going to stand up and *fight* for their land."

That fight ultimately proved successful: a moratorium was placed on the cutting of eighteen thousand acres of public land, the conversion of Bankhead Forest into commercial pine plantations was halted, and a number of sacred sites have remained untouched ever since.

Looking back, Marshall said, his entire life could be seen as a preparation for the wildly multidisciplinary work of mapping ancient trails. The years of hiking taught him to navigate cross-country, the career in trapping taught him about lines of habitual movement, and his lifelong study of American Indian cultures taught him why a Native trail might go somewhere a European trail would not. Working on surveying crews taught him how to read geographic surveys and draw maps. And years spent hunting, fishing, attending Baptist church services, and cussing liberal bureaucrats—in short, living the life of a red-blooded son of the Southland—allowed him to talk with hillbillies and Cherokee elders alike, gathering information a desk-bound academic might otherwise miss.

+

In the end, Marshall concluded that the trail we'd walked that November morning was among the best preserved Cherokee trails in western North Carolina. He later found mention of it in an account written by an army captain named W. G. Williams, who led a secret reconnaissance mission into Cherokee country in 1837 in preparation for the infamous Cherokee Removal. (Williams described it, tersely, as a "very rugged trail.")

Initially, Marshall had dubbed it the "Big Stamp Trail," because it eventually climbed its way to a high grassy summit called Big Stamp. A "stamp," in local vernacular, is a place where large numbers of deer (or formerly, buffalo) gather to graze or access salt licks, stomping down vegetation in the process.

There was some disagreement, however, over the name. On the morning of our hike Marshall had met a bear hunter named Jimmy Russell, who corrected him when Marshall said he was hiking up toward Big Stamp. "We call it Big *Stomp*," Russell said. Marshall scribbled this down in his notebook.

A few hours later, Marshall received a call on his cell phone from his neighbor, Randy, who happened to be another bear hunter. (Bear hunters, Marshall explained, were an excellent and underutilized intellectual resource—because they had to scramble cross-country in pursuit of their quarry, they knew every trail in the mountains, even the abandoned ones.) Marshall told Randy we were surveying the trail up to "Big Stomp."

Randy interrupted him. "We call it Big *Stamp*," he said.

"Well, the map calls it Big Stamp, but we were corrected," said Marshall. "That mountain guy told us they call it Big *Stomp*."

"Aww, well we always call it Big *Stamp*. That's what *we* call it," Randy said.

Marshall made a note of this, too. (He ultimately stuck with "Stamp.")

In this line of work, names matter. In the absence of reliable maps, Marshall was often forced to stitch together prospective paths from the town names he found in written records. This task was made exponentially more difficult by the fact that explorers and surveyors tended to err badly (and often, bizarrely) when transliterating Cherokee place names. For example, the Cherokee village of Ayoree Town was inaccurately renamed Ioree, which then became the Iotla Valley. George R. Stewart pointed out in his masterful *Names on the Land* that Europeans, accustomed "to names like Cadiz and Bristol which had long since lost literal meaning," were often content to mangle Native Americans' highly descriptive, intricately wrought place names, using them as mere tags, much like how an archaeologist might use a Stone-Age knife as a paperweight.

Sometimes when Marshall was stuck on a curious place name, he would take it to a Cherokee linguist named Tom Belt, who could decode it. Not long ago, for example, Belt informed him that Guinekelokee (what is now the West Fork of the Chatooga River) meant, "where the trees hang over the sides."

I visited Belt one afternoon at his office at Western Carolina University. He wore cowboy boots, blue jeans, and a silver belt buckle. Around his neck, over a purple dress shirt, hung an abalone pendant engraved with woodpecker heads. A mop of gray-streaked hair was cut just above his boyish eyes. His voice had a warm, dark, smoke-rasped, far-off quality.

Belt was born and raised in Tahlequah, Oklahoma. His ancestors were brought there in the decade following the signing of the widely reviled Indian Removal Act of 1830 by President Andrew Jackson. (Belt's feelings about Jackson were plain: on the wall of his office hung a photo of Jackson that had been fashioned into a WANTED poster.) Some of the Cherokee had gone west peacefully, but many were moved by force, shuttled under armed guard along what they called *Nvnohi Dunatlohilvhi*, "The Trail Where They Cried," or as it is more commonly known, the Trail of Tears. Sixteen thousand Cherokees were driven from their homes; while some were carried on riverboats, others were forced to walk almost a thousand miles across inhospitable country. Perhaps as many as four thousand men, women, and children died, mostly of disease.

The full ramifications of the Removal, and the pain it inflicted, are difficult for non–Native Americans to grasp. As Belt made clear to me, our two cultures have a drastically different "sense of place." To Euro-Americans, places are most often regarded as sites of residence or economic activity—essentially blank backdrops for human enterprise. As such, Euro-American places are largely ahistorical, *replaceable*; they change hands, and their names can change too. By comparison, the Cherokee conception of place is more fixed, specified, eternal. "In the native world, places don't change identity," Belt said. "We are more in touch with place as where things *have happened*, and where things *are*, as opposed to where *we* are."

The Cherokee derived their tribal identity from an ancient townsite twenty-five miles west of where we were sitting, called Kituwah, "the soil that belongs to the creator." Cherokee villages once ranged across the Southeast, from Kentucky to Georgia, but, according to Belt, if you had asked any of those villagers where they were from, they would have told you they were *Otsigiduwagi*, the people of Kituwah. On a ceremonial mound in the mother town burned an eternal flame. Once a year, coals from that flame were carried to the various towns. Thus was the vast Cherokee nation strung together: by language, by narrative, by ancestry, by tradition, by glowing embers carried along a network of trails.[*]

There are myriad reasons—historical, cultural, and economic—why the Anglo-American sense of place diverged so radically from that of the Cherokee. However, Belt believed that a crucial, often overlooked difference lay in the very structure of the respective languages. The Cherokee language differs from English in key ways. Cherokee has seven cardinal directions that continually situate speakers in space: *north, south, east, west, up, down*, and (hardest of all for us outsiders to grasp) *here*. The structure of Cherokee grammar—in which the subject of a sentence comes after the direct object—also serves to subtly decenter the speaker. "In the English language it's *I think this, I think that; I want this, I want that*. It's as if we're in the

---

[*]When the Cherokees were forced to move to Oklahoma beginning in 1830, embers from that eternal flame were carried west, and the fire was rekindled there. Then, in 1951, the flame was brought back to North Carolina, where it now burns in front of the Museum of the Cherokee Indian in the town of Cherokee.

center of the world and the world is around us," Belt said. "In our language, everything is *here* and we're some place around it. Which means that we're just a part of it, as opposed to being in the center of it." Moreover, Belt noticed, the Cherokee word order was better suited to a wild environment. As he pointed out, when a bear is sneaking up on your friend, it helps for the sequence of words coming out of your mouth to be "bear . . . I . . . see" rather than "I . . . see . . . a . . . bear."

Belt's upbringing made him acutely aware of the ties between geography and language. As a boy, he spoke only Cherokee; he didn't learn English until he was seven years old. Back then he spent his afternoons playing war games with his friends in the prairies of Oklahoma. In his mind, however, he always fantasized that he was in a land of mountain slopes, soaring trees, and murmuring brooks. When he moved to North Carolina, at the age of forty, he was shocked to realize that this was the landscape he had always been imagining. A friend of his, who grew up not far from where he did, recounted a similar experience. She showed him a picture she'd drawn when she was five or six. The terrain of the background was verdurous, mountainous, utterly unlike anything she'd ever seen in Oklahoma. The same landscape appeared in the background of all her drawings, she said.

"It wasn't until she came here that she realized what she was drawing," Belt said. "She was drawing these mountains."

This sense of deep geographic memory may seem mystical, he said, but it isn't—or at least, isn't entirely—because the landscape is "encoded" directly into the language. Cherokee diction and syntax are mountainous. The language has several fine-grained descriptors for different types of hills. Suffixes can be appended to nouns to indicate whether an object is uphill or downhill from the speaker. (If there is a river nearby, objects can also be described as upstream or downstream.) In the flatlands of Oklahoma, this mode of description seemed odd to Belt, until he came to the mountains of North Carolina, and then it made perfect sense.

+

Barbara Duncan, a folklorist who has spent decades recording Cherokee myths and legends, told me that she had noticed a curious difference between the eastern and western halves of the Cherokee nation. The stories of the eastern Cherokee, those who avoided the Removal, are often more geographically rooted than those of the western Cherokee, she said. She cited an ancient folktale about a race between a turtle and a rabbit, in which the clever turtle fools the cocky rabbit by positioning his brethren on top of a series of peaks, so that every time the rabbit crested one mountain, he was shocked to find the turtle ahead on the next. The recollections of eastern Cherokees mentioned that the story occurred on what is today called Mount Mitchell, whereas those of western Cherokees typically do not specify a location. "And if you go to Mount Mitchell, you can see the land formation that is described in the story," Duncan said. "You can tell the story without ever going to Mount Mitchell, it's still an entertaining story. But when you go up on top of that mountain and you see that landform, you're like 'Oh, *this* is what they're describing.' It's amazing."

"Almost every prominent rock and mountain, every deep bend in the river, in the old Cherokee country has its accompanying legend," noted the ethnographer James Mooney. "It may be a little story that can be told in a paragraph, to account for some natural feature, or it may be one chapter of a myth that has its sequel in a mountain a hundred miles away." This phenomenon, Mooney wrote, extended well beyond the Cherokee. In the storytelling traditions of virtually every indigenous culture, stories don't unfold abstractly, like Little Red Riding Hood skipping through unnamed woods; they *take place*. The stories of the Inuit, for example, always specify a real setting where the story (often, a depiction of a journey) unfolds; many stories even include details about the direction of the prevailing wind.

In his landmark study of the Western Apache, *Wisdom Sits in Places*, the linguistic anthropologist Keith Basso limned the many ways that land and language help construct indigenous cultures. First, places were named, often in intricate visual detail ("Water Flows Inward Under a Cottonwood Tree," "White Rocks Lie Above in a Compact Cluster"). Once named, those places became what Basso called "mnemonic pegs" to which stories—creation myths, morality tales, ancestral history—were attached and group identities were formed.

Apaches view the past as a well-worn trail (*'intin*), once traveled by their ancestors, and still being traveled today. "Beyond the memories of living persons, this path is no longer visible," wrote Basso. "For this reason, the past must be constructed—which is to say, imagined—with the aid of historical materials." Apaches relate this process of re-creation to how one can reconstruct a person's movements from scattered footprints. Time frames grow vague, and characters are often reduced to archetypes, but the essential elements—the settings, the lessons, the flora and fauna—remain highly specific. ("Long ago, right there at that place, there were two beautiful girls . . ." begins a typical story.) Basso notes: "What matters most to Apaches is *where* events occurred, not when, and what they serve to reveal about the development and character of Apache social life."

In a delightful twist, Basso's work also provided a mirror view of just how strange the prevalent mode of Euro-American storytelling is. Upon hearing European stories read aloud to them, many Apaches told Basso they found them as inert as the paper on which they were written. By comparison, Apache oral narratives were vivid, fluid; they shifted subtly with each telling, in accordance with the whims of the speaker and the disposition of the listener. Apache stories may not have been strictly accurate by academic standards, but they were wise, witty, and most important, they *worked*. To teach someone a lesson, Apache elders would often tell that person a story about a specific place. For example, a careless boy might be told the story of the canyon where a girl took a shortcut against her mother's instructions and ended up getting bitten by a snake. That way, every time the careless boy passed by or even heard mention of that canyon, he would be reminded of the lesson. It was, therefore, no exaggeration when Apaches said that a place "stalks" them, or that the land "makes the people live right."

In Apache culture, places do not exist in isolation. Rather, as in nearly all indigenous cultures, places are linked together in a spatial and conceptual matrix, flowing one to the next. On one occasion, Basso noticed an old Apache cowboy talking quietly to himself. When Basso listened carefully, he learned that the old man was reciting the names of places, one after another—"a long list, punctuated only by spurts of tobacco juice, that went on for nearly ten minutes."

Basso asked him what he was doing, and the old cowboy replied that he "talked names" all the time.

"Why?" Basso asked.

"I like to," the old cowboy replied. "I ride that way in my mind."

Anthropologists have a term for this practice of place-listing: topogeny. It is storytelling at its most spare, rendering a narrative down to a string of dense linguistic packets, like seeds, which flower in the mind. It has been observed in locations as far-flung as Alaska, Papua New Guinea, Vancouver Island, Indonesia, and the Philippines. The list of names serves to pull the mind across the landscape—from mnemonic peg to mnemonic peg, from story to story—following a geographic line. According to the anthropologist Thomas Maschio, the Rauto tribe of Papua New Guinea could recite hundreds of place names in a row. "To remember the names of these sites, elders said that they 'had to walk' along the various paths," Maschio wrote. "As I sat with the elders in the men's ceremonial house, the sequence of place names was recited to me as if the elders were taking part in a journey or imaginary walk through the many paths of the land. Elders would name a place, tell me its history, and then say that they would now 'walk on to the next place.'"

Topogeny is not simply the listing of names; it is the summoning, in the mind's eye, of a mental landscape constructed of lines. This notion struck me one day the following summer, when I went on a hike with Lamar Marshall alongside Brush Creek, near the old Cherokee town of Alijoy, an hour's drive from Asheville. From time to time along our walk, he paused to gather plants the Cherokees living there would have found useful: a fragrant pinch of spicebush, a handful of fibrous bear grass, a bright yellow knot of medicinal goldenseal root. On the banks of the creek, he spotted a beaver slide, and showed me how he would have once set a trap there.

Though he moved fluidly through the thick switch cane, Marshall was having trouble catching his breath. He told me he spent too much time inside, staring at old maps and documents. His passion for research was beginning to border on obsession.

"My wife is on me about it right now," he said. "We're in the middle of one of the greatest places in America: an inordinate amount of trails, scenic beauty, rivers. And how many times have I fished this year? Only once. I've only been in my canoe for four hours. Every year I say, 'This year is going to be different. I'm gonna fish, I'm gonna hike, I'm gonna backpack, I'm gonna camp.' And then the year gets gone, and it's like, 'I just turned sixty-six!'"

But all that time spent inside studying old maps and stories seems to have only strengthened his connection to the land, oddly enough. In the six short years since he moved to North Carolina, his knowledge of the history and geography of the region had grown truly encyclopedic. The most striking thing, I noticed, was *how* he spoke about history: his recollections were almost always structured spatially, rather than chronologically. For him, as for that Apache cowboy and those Rauto elders, the land was furnished with hundreds of mnemonic pegs.

"People are amazed because I can draw a map of all of western North Carolina," he said. "I can draw all the watersheds. I can put in probably close to sixty Cherokee towns. And it's not like I've got a list in my mind, I've memorized A, B, C, D. No, I'm *visualizing* the trail going up over Rabun Gap, down into the upper branches of the Tennessee River . . . My mind just flows over the mountains, down the valleys, along the trails, through the thickets . . ."

He closed his eyes and tilted his head back, seeing something I could not.

> "There's: Estatoe Old Town,
> Kewoche Town,
> Tessentee Town . . .
> Skeena Town,
> Echoy Town,
> Tassee Town . . .
> Nikwasi.
> Cartoogechaye.
> Nowee.
> Watauga.
> Ayoree.
> Cowee.
> Usarla.
> Cowitchee.
> Alijoy.
> Alarka . . ."

+

Marshall had brought me out to Brush Creek that morning so we could look at a stretch of previously undiscovered wagon road that was part of the Trail of Tears. He was fighting to gain federal protection for it as a historic place—its was a dark history, but an instructive one nonetheless. It was rare to find an undeveloped stretch of the Trail of Tears. Much of the rest had already been assimilated into the modern road network.

The Trail of Tears was far from a unitary trail. What we call the "trail" was in fact a spider-veined array of paths along which tribes were transported, including a number of river routes. In 1987, President Reagan designated certain stretches of that network a National Historic Trail, to memorialize the one hundred fiftieth anniversary of the Removal. Every year, some one hundred thousand motorcyclists ride one of its legs—now a series of highways—west from Chattanooga, Tennessee to Waterloo, Alabama, in solidarity with the removed tribes.

We got out of the car and crossed a swinging bridge across Brush Creek, then we walked down a gravel road until we reached a tributary, where we tiptoed over a log with wooden footholds nailed to it. ("Redneck Bridge," Marshall chuckled.) On the other side of the creek ran the forgotten stretch of the Trail of Tears. It was flooded with dark water. Otherwise, it was remarkably well preserved. The passage of countless wagons had cut a wide, muddy runway. Before it had been a wagon road, it too would have been a Cherokee footpath.

"You can follow this for miles," Marshall said, looking off at where it disappeared into the trees.

Standing there, the cruel irony of not just the Trail of Tears, but all Native trails, hit home. Over the course of thousands of years, Native Americans devised a beautifully functional network of paths, not knowing that those same trails would later be used by a foreign empire in its slow invasion. Along their trails flowed surveyors, missionaries, farmers, and soldiers, as well as diseases, technology, and ideology. Then, when a critical mass of foreigners had moved into tribal lands, it was along those trails that Native families were hauled from their home. We tend to think of colonialism as an unstoppable wave, or a platoon of tanks moving smoothly across the plains, when in fact it is more like the trickle of an ever-multiplying virus through an arterial network.

From the very beginning, Europeans exploited Native wisdom,

Native kindness, and Native infrastructure. Across the continent, many of the easiest mountain passes were discovered only when Native American guides led white men there. Henry Schoolcraft located the source of the Mississippi due only to the guidance of an Ojibwa chief. Following his Native guides across Baja California, an explorer named James Ohio Pattie slept on the edge of their blankets at night so they could not sneak off without waking him. (If they had escaped, he wrote, "we should all undoubtedly have perished.")

One of history's most striking examples of how Europeans relied on local wisdom was provided by a rather mysterious figure named John Lederer. Almost nothing is known about Lederer prior to his arrival in Jamestown in the 1660s, save the fact that he was a doctor from Hamburg. Though he spoke little English, he did speak fluent German, French, Italian, and Latin (the language in which he later recorded his travels). Obstinate and ambitious, he was prone to amassing huge debts and bitter enemies. It is unclear how such a figure convinced the governor of Virginia, William Berkeley, to appoint him to search for a passage through the Appalachian Mountains to the West. But by March 1669, perhaps less than a year or two after arriving in the New World, he had already embarked on his first expedition. He hired three Native American guides to lead him. The trip was arduous—along the way, he was almost swallowed by quicksand and feared his horses would be devoured by wolves—but five days later, his party reached the foothills of the "Apalataean mountains." When Lederer first glimpsed the famous range, he could not decide whether they were mountains or clouds, until his guides fell to their knees in prostration and howled out a phrase meaning "God is nigh." (Or so he claimed.) Reaching the range, he attempted to ride up to the top of the mountains on his horse, but the horse balked. Lederer's guides were clearly also unfamiliar with the region, because they next attempted to scale the mountain on foot, without the aid of a trail, and soon became entangled in brush and brambles. The route was so steep in places that, when Lederer looked down, his head swam. Though they had set out at first light, it was nightfall again before Lederer and his companions finally summited. They camped amid the dark boulders. He awoke high on the mountain the following dawn and looked west, expecting to find the sparkling waters of the Indian Ocean. Instead, there was only a wall of yet taller mountains. Hoping to find some passage through the range, he wandered among those snowy peaks for six days, drinking from springs where the water tasted faintly of aluminum, his hands and feet growing numb in the thick, chill air, before he gave up and returned home.

On a second expedition in May 1670, Lederer planned a more ambitious assault, leaving with five Native American guides and twenty Englishmen under the command of one Major William Harris. Lederer had wised up during his previous expedition. The most important lesson he had learned was that one could not navigate the mountains without the help of local tribes, who knew the easiest routes through the mountains. He took care to bring along a store of trade goods to win their confidence: sturdy cloth, sharp-edged tools, dazzling trinkets, and strong liquor. He had also learned how to travel in comfort over the strange continent. At night, instead of sleeping on a bedroll, he slept in a hammock, which was "more cool and pleasant than any bed whatsoever." He fed himself by hunting for deer, turkeys, pigeons, partridges, and pheasants; when he was nearing mountains where game would be scarce, he prepared a pile of smoked meat in advance. Instead of biscuits, he brought along dried corn meal ("i.e. Indian wheat"), which he seasoned with a pinch of salt. The Englishmen laughed at his odd food, until their biscuits turned moldy in the humid air. Then they tried to beg Lederer's corn meal from him, but—"being determined to go upon further discoveries"—he refused to share.

Two days after setting out, the party reached a village marked by a pyramid of stones. They asked the villagers for directions to the mountains. An old man obliged to describe the route for them, drawing a map in the dirt with his staff. He depicted two paths that meandered through the mountains, one north and the other south. However, Lederer's companions, slighting the old man's advice, decided to strike their own course, following their compasses due west. Reluctantly, Lederer tagged along. For the next nine days, the party exhausted their horses by riding over rough terrain and scaling craggy cliffs. Lederer compared their progress to that of a land crab, which, crawling up and over every plant in its path rather than circumventing them, covers less than two feet of ground after a day's hard labor. The expedition finally reached a river flowing north, which Major Harris (perhaps willfully) mistook to be an arm of the fabled "lake of Canada." Having found this important landmark, Harris decided the party should return to Jamestown. Lederer disagreed, and an argument ensued. Harris's men, starved and exhausted, threatened Lederer with violence, but he staved them off with a letter from the governor granting him permission to push forward. Harris and his men turned back, leaving Lederer with a gun and a single guide, a Susquehannock named Jackzetavon. Harris returned to the colony and, according to Lederer, began to "report strange things in his own praise and my disparagement, presuming I would never appear to disprove him."

Lederer and Jackzetavon pushed on, traveling from village to village, stopping frequently to ask the local chieftains for directions. The information Lederer gathered, much of it refracted through an unknown number of translations, often shimmers with the exotic air of a colonial fantasy, replete with human sacrifices and temples full of pearls. Some of the tribes he encountered were tall, warlike, and rich; others were lazy or effeminate. Some were governed by democracies, others by ruthless monarchies, while others still held everything in common ownership ("except their wives," Lederer primly added). In one village, Lederer watched a man step barefoot onto a bed of burning coals and stand, writhing, foam collecting on his lips, for a full hour, before leaping out, apparently uninjured. At times, Lederer was impressed by the Native peoples' resourcefulness—as when he witnessed the delicate process by which acorns were roasted and pressed to yield an amber-colored oil, which, sopped up with corn bread, provided "an extraordinary dainty." Yet he was also horrified by what he perceived as the tribes' gleeful love of violence, as when a group of young warriors returned from a raid to proudly present their chief with "skins torn off the heads and faces of three young girls."

Despite the Swiftian tone of his writing, in contrast to the other explorers of his age, Lederer comes across as relatively peaceable, respectful, and punctilious. His second expedition lasted for some thirty days, and he made it home safely. Shortly afterward, he set out on a third, failed expedition in which, after only six days, he was bitten on the shoulder by a deadly spider and only managed to survive thanks to a Native American man who sucked out the poison.

Back home in Maryland, Lederer compiled his notes into a narrative and drew a map showing the route of his journeys. For a time, his writings were dismissed as too fantastic to be believed, but scholars have since judged them to be surprisingly accurate, given the obvious technical limitations of the time. (One must keep in mind just how little white people then knew about the extent of the American continent. Many then believed the Indian Ocean lay only one or two hundred miles west of the eastern seashore, a theory Lederer roundly debunked.) Lederer's account included a wealth of information, including the location of two easily navigable passes over the mountains, both of which were confirmed by explorers in the following years. One of those passes was described for him by an unnamed group of Native Americans, the other he had seen while following Jackzetavon. It is possible they were the same two paths the old man had drawn in the dirt, which Major Harris, in his arrogance, ignored.

The example set by Lederer and Harris was repeated almost everywhere that the colonists spread: those who ignored the advice of Native people and spurned their trails ended up tangled in brambles and mired in swamps, whereas those who co-opted Native wisdom moved smoothly. A century after Lederer, by following several well-known Native trails, the famed mountaineer Daniel Boone and a team of thirty-five loggers would cut a horse trail up and over the Appalachians through the Cumberland Gap, opening up the continent to westward colonial expansion.

+

Native guides were sometimes called "pathfinders," a title that has a double meaning: in wild and remote areas, their job was indeed to locate obscure trails, but in more densely populated areas—which were at times so thickly webbed with trails that explorers described them as a "maze"—the pathfinder's job was to chart a course through that network. A scholar in North Carolina told me that he had recently read a history of Hillsborough that began with the vague, romantic-sounding claim that when Hillsborough was founded, the county was a "trackless wilderness." "That's such bullshit!" he exclaimed. "The problem wasn't that it was trackless; it had *too many tracks.* That's why you needed a Native American guide—to tell you which one of these roads to use."

In the absence of a trusty guide, the best alternative for navigating a path network is to use a system of signs. Long before the painted blazes that demarcate modern hiking paths, people were slashing blazes into tree trunks to mark their trails. Across the continent, tree trunks were also marked with bright paint, elaborate carvings, or sketches drawn using a mixture of bear fat and charcoal. Along snowshoe paths in northeastern Ontario, evergreen boughs were inserted into the snow at regular intervals to serve as signposts. In many places—from Montana to Bolivia—stones were piled up into cairns, which served both a functional and a spiritual purpose. When I was herding sheep in the Navajo country, I often ran across large stone cairns, which I was told were used in prayers, but which also helped us herders find our way home.

Perhaps more common, but certainly more difficult to document, were the subtler trail signs indigenous people left in passing. For instance, the practice of using bent or broken twigs to mark trails and transmit messages is widespread. In his supremely ill-titled account of transcontinental exploration, *First Man West*, Alexander Mackenzie wrote that his Native guides marked the trail "by breaking the branches of trees as they passed." Elephant hunters in Africa noted that it was customary to mark one's trail by blocking tributary trails with a stick laid across the offshoot, as if closing a gate. The Rauto tribe of Papua New Guinea have a word, *nakalang*, for the stick that bars an errant trail. In their language, the word is also used, poignantly, to signify death, which separates the divergent paths of the deceased from those of the living.

A few years ago, I went for a walk through the Bornean jungle with Henneson Bujang, a Penan tribesman, and his two sons. As we walked down faint footpaths, they showed me how to bend or break a twig to send a message to someone down the trail. Bujang estimated that he knew dozens of signals involving broken twigs for sending messages, which could be as specific as "Avoid this trail—there's a hornet's nest here," or "You're taking too long, I'm hiking up ahead." The author Thom Henley recorded a number of stick signs from the groups he visited: A four-pronged stick planted in the ground indicated a burial, whereas three sticks arranged upright, like a fan, marked a territorial claim. One particularly elaborate stick sign he found in the Melinau River drainage showed the richness of information that could be encoded in these signs:

A large leaf at the top showed that the stick had been left by the headman. Three small uprooted seedlings indicated that the site had once been occupied by three families. A folded leaf told that the group was hungry, in search of game. Knotted rattan gave the number of days anticipated in the journey and two small sticks equal in length and placed transversely on the sign stick suggested that there was something for all Penan to share. Sticks and shavings at the base identified the group and revealed the direction of the journey.

Life is a continual struggle to make sense of the world's complexity. Knowledge is hard won, and so both spoken language and writing are ways of fixing and transmitting it. Though we tend to imagine that there is a sharp dichotomy between oral cultures and those that have developed written language, as trail signs reveal, there is a vast array of media—twigs, cairns, drawings, maps—that blur the line between the two. But perhaps the simplest and yet most dissolvent of all sign systems, the ur-inscription that predated writing and even the spoken word, is the trail itself.

+

Henneson Bujang and Lamar Marshall had something in common. While Marshall was fighting clearcutting in Bankhead Forest in Alabama, Bujang and a handful of other Penan tribesmen began sabotaging logging companies' attempts to cut down the old-growth rainforest where they live. Against steep odds and immense pressure, both ultimately succeeded.

Though some truly nomadic, hunting-and-gathering members of the Penan still lived in the hinterlands of Borneo, the Bujangs had since settled into the comforts of Christianity, zinc roofs, and shotgun ownership. But they continued to doggedly oppose logging efforts (and refused considerable bribes) for reasons that were as much coldly practical as they were culturally inherited: almost all their food still came from the jungle. Over the two days I spent with them, we ate wild bearded pig, mouse-deer, fish, small birds, ferns, local rice, green chilies, and cucumbers—all of it collected from within three miles of where we sat. The birds were hunted using a blowpipe, which Bujang wielded with a sniper's accuracy, hitting hornbills in treetops up to two hundred feet away. He also knew how to carve a bowl out of ironwood, weave a rattan basket, and build a bamboo shelter that would keep him both dry and elevated off the jungle floor. On our walks, he and his sons showed me which insects were edible, which leaves were antiseptic, which plants would cure a headache, and which giant ferns could be woven together to form an impromptu umbrella. "The earthworm can go hungry and the mouse-deer become lost in the forest, but never we Penan," the Penan like to say.

From the time they have grown waist-high, Penan children are traditionally taken on long journeys through the jungle to learn the land. Even just one generation ago, Penan parents would pull their children out of school for weeks or even months at a time to teach them the old stories and skills. But in recent years, Penan children have become too busy with schoolwork to memorize, preserve, and pass down this ancestral knowledge. Talking with Bujang, I came to realize that the most pernicious threat to their culture might not be the logging companies themselves (which can be fought by blocking roads and "spiking" trees); it's the slow but unceasing creep of the loggers' worldview.

Bujang worried over this cultural erosion one afternoon, as we sat around his dinner table, eating stubby blackened bananas. His sons were sprawled on the floor, playing with a skittery baby macaque and a lazy, bug-eyed pangolin they had taken as pets. Bujang said that while his sons were decent hunters, they lacked the instinctive directional sense that allowed him to walk for miles through dense, pathless brush without getting lost—a skill that only comes from a lifetime spent in the jungle. He worried that his grandsons and great-grandsons would surely only be weaker, less knowledgeable, and more dependent on the state for their survival.

We who live in industrialized societies sometimes tacitly assume that all hunter-gatherers will inevitably want to "graduate" to practicing agriculture and global capitalism, but as Bujang and many other indigenous families prove, the so-called modern world is not humankind's manifest destiny. Some hunter-gatherer societies, like the Cheyenne, chose to abandon a lifestyle of sedentary farming and return to hunting and gathering. Many others, like the Bujangs, chose to adopt certain elements of Western modernity but not others: they hunt bush pig with a shotgun, but kill birds with blow darts. The path of humanity is ever branching. All roads need not lead to Times Square.

However, as I learned from the Bujangs, the path of modern consumer capitalism—with its endless (and endlessly advertised) comforts and conveniences, its wondrous medicine and magical technology—constantly beckons to hunting-and-gathering societies. Once Western capitalism begins to encroach on their land from all sides, life for hunter-gatherers becomes increasingly difficult—their land base shrinks, government interference increases, traditions fray, local knowledge dissipates, and the pressure to assimilate mounts. Under these conditions, converting to a sedentary Western lifestyle begins to look like the path of least resistance.

When an indigenous community assimilates into the dominant culture, either by force, by desire, or from fears of being "left behind" (a threat the Malaysian government frequently wields), there is a concomitant loss of those threads that hold together culture: language, lore, religious practice, familial obligations, and relationship to place. At its core, the problem facing indigenous communities is mnemonic; the culture, long stored in the collective memory and encoded in the land, is gradually forgotten.

Modern Cherokees hold many of these same concerns, but they

have their own way of fighting the erosion. Unlike the Bujangs, or traditional Navajos like Bessie and Harry Begay—who have retained much of their culture by living far from civilization, shunning most technology, and never learning English—the Cherokees have tried to chart a middle path between assimilation and traditionalism. Virtually from the beginning of the imperial invasion, many Cherokees were quick to learn the English language, utilize modern technology, and adopt European modes of farming and trade, all while fighting to maintain their heritage.

That fight continues to this day. Since language is vital to preserving culture, in recent years the Cherokee—like many other tribes—have founded a slew of language-immersion schools, which help children become fluent in their native language, even if their parents can't speak it. I paid a visit to one such school, the Kituwah Academy in North Carolina. The school was run by a man named Gilliam Jackson. Out of roughly fourteen thousand Eastern Band Cherokees, Jackson was one of only a few hundred left who still spoke the Cherokee language fluently.

At Kituwah, every activity—classes, games, meals, songs—was conducted in Cherokee. Over the entryway to the school hung a printed banner: ENGLISH STOPS HERE. In one classroom sat a pile of colorful wooden alphabet blocks, such as one might find in any preschool, only, instead of the Latin alphabet, they were printed with odd-looking characters that resembled something dreamed up by Tolkien. This, Jackson explained, was the Cherokee alphabet (properly known as the "Cherokee syllabary"). Most fluent Cherokee speakers cannot read or write it, but these children were learning to do both.

The written language of the Cherokee was devised in the early nineteenth century by a Cherokee blacksmith who went by the name of Sequoyah, also known as George Gist. Unable to speak English, Sequoyah marveled at the efficiency of written language, which allowed white people to converse across great distances and, most crucially, to fix knowledge so that it would not erode over time. In the oral tradition, knowledge was a mercurial thing, changing shape as it changed hands. As reported in *The Missionary Herald* in 1828, Sequoyah believed that, "if he could make things fast on paper, it would be like catching a wild animal and taming it."

Seeing the benefits of such a technology, he set about devising his own code. He began by assigning individual symbols to common words. First he used elaborate hieroglyphs, which, being tedious to draw and memorize, he later replaced with more basic symbols. But with his list of words piling up into the thousands, even those symbols soon proved too difficult to remember. Finally, after much experimentation, he broke the language down to eighty-six spoken syllables, with a different character assigned to each. It took him twelve years to find a workable system.

Once the system was complete, he taught it to his six-year-old daughter, who came to read it fluently. Together, they demonstrated the new system for their neighbors: Sequoyah would ask someone to tell him a secret, he would mark it down on a piece of paper, and then he would ask another person to carry the paper to his daughter, who was out of earshot. Reading from the scrap of paper, the little girl would call out the sentence just as it had been told to her father, shocking many of those present.

Sequoyah's new syllabary caught on. It was soon being used to record sacred songs and curing formulas. By 1828, a new bilingual newspaper called the *Cherokee Phoenix* was published in an altered form of the syllabary. In the 1980s, the first Cherokee typewriter was invented, and attempts were later made to convert computer key-boards into Cherokee. But, due to mechanical and cost constraints, typing in Sequoyan remained cumbersome until 2009, when a developer released an application for electronic tablets like the iPad that allowed people to easily type in Cherokee.

Jackson said the children at his school had picked up typing in Cherokee on the tablets with remarkable ease. "We have really gotten way into technology," he said. "All these kids around here, they can text, they can do iPads, they can do computers, you name it. They're way more advanced than I am. But in terms of being able to identify plants and medicines and foods in the woods, they've lost that connection."

As Jackson was learning, the cultural institutions that European cultures have long relied on to perpetuate knowledge—namely, enormous and intricately organized corpora of texts—cannot properly preserve a form of knowledge that is orally transmitted and terrestrially encoded. Indigenous cultures need both language and land to survive.

+

People fighting to preserve indigenous cultures tend to fall into one of two camps. Some believe that technology (being malleable and agnostic) will continue to evolve to better perpetuate elements of indigenous culture, like the Cherokee keyboard, and to situate traditional knowledge in the landscape (using digital maps). Others, like Jackson, counter that without time spent learning directly from the land, no amount of technology would halt the cultural erosion.

Somewhat ironically, given his general aversion to technology, Lamar Marshall had ultimately been converted by the techno-evangelists. In response to the loss of land-based learning, he has begun importing over a thousand miles of trails into digital maps—along with the stories, wild foods, and medicine to be found along those trails—so they could one day be accessed by future generations of Cherokees.

The day after our trip up to Big Stamp, Marshall met me at the Wild South office to show me an early mock-up of the program. Using Google Earth, he had charted a shortcut trail that connected the Raven Fork Trail to the Soco Creek Trail. The satellite images were from the present day, so occasionally, as the yellow path wended through the green mountains, the gray specter of an asphalt parking lot would appear. The anachronisms had initially jarred Marshall, and he considered using Photoshop to clone trees in over the modern scars, but he ultimately decided against it. The program, he reasoned, should represent the world as it currently is—or rather, as it can be walked.

Into these digital landscapes, here and there Marshall had inserted images that the ancient Cherokee would have found alongside the trail: ginseng leaves, elk, bison, a trail marker tree, intersections with old trading paths. On one hill, called Rattlesnake Mountain, lurked a crude rendering of Uktena, the horned serpent of Cherokee myth.

As it was told to James Mooney, the monster Uktena was known to hide in dark, lonesome passes over the Great Smoky Mountains. One day a local medicine man named Aganunitsi went hunting for the great serpent, hoping to collect the diamond that was embedded in its forehead. He walked south through the Cherokee lands, encountering mythic snakes, frogs, and lizards, before he finally reached the top of Gahuti Mountain, where he found the Uktena sleeping. The medicine man retreated to the bottom of the mountain and made a great circle of pinecones. Inside the circle he dug a deep circular trench, and within that, he left an island on which he could stand. Then the medicine man lit the pinecones on fire, crept back up to the serpent, and shot an arrow through its heart. The snake awoke in a fury and lunged after the medicine man. The man was prepared; he ran down the hill and leaped inside the circle of flames. The snake raced down behind him, spraying venom, but the poison evaporated in the blaze. While the man waited safely inside his ring of fire, the injured serpent roiled in agony, flattening trees. Its black blood poured down the slope, filling up the circular trench. The medicine man waited on his little island until finally the beast fell limp. After waiting seven days, he visited the site where the serpent lay, and though its flesh and bones had been pecked to dust by birds, one thing remained: a luminous diamond. With that jewel, the medicine man soon became the most powerful man in the tribe.

Encoded within the story of Uktena—which I have greatly abridged here—is an enormous amount of information, both empirical and mythic, all spun into a taut narrative thread. Even on the page, the story hums. One can only imagine how much more vivid it would have seemed if one had heard it while standing on the mountainside, looking out over the treeless expanses swept bare by the serpent's tail and the lake filled with its black blood.

Marshall's program was a small but meaningful attempt to resituate the story in its rightful place. However, it still lacked the immediacy of terra firma. Marshall knew this, so he hoped to one day build an application that incorporated augmented reality technology with stories and maps, so that children could stand on the slopes of Rattlesnake Mountain while watching the tale of the Uktena unfold through virtual-reality goggles, or visit the sacred Kituwah mound and see a digital rendering of the site as it once was, four centuries earlier, aglow with the light of the sacred fire.

+

Walking creates trails. Trails, in turn, shape landscapes. And, over time, landscapes come to serve as archives of communal knowledge and symbolic meaning. In this sense, the various cultures I have so far crudely lumped together under terms like *Native* and *indigenous* could perhaps be better described as "trail-walking cultures." This classification would make modern Westerners, by extension, a "road-driving culture." The colonization of the New World would not have been possible if Europeans could not harness domesticated animals and drive in vehicles like wagons and, later, trains and automobiles. Today, machines allow us to move radically faster—often along the very

same trails Native Americans once used. But in doing so, we have
lost the elemental bond between foot and earth.[*]

The Blackfoot Indians of North America are an archetypal example of a trail-walking culture. According to their creation stories, the
world was shaped by a quasi-divine figure named Napi, or Old Man,
as he walked north through Blackfoot country. In the process, he
formed rivers, planted deposits of red clay, gave birth to animals. He
created plants to feed the animals. Then he created humans to hunt
the animals and harvest the plants. He showed humans how to dig up
edible roots, how to gather medicinal herbs, how to hunt with a bow
and arrow, how to drive buffalo over a cliff, how to use a stone maul,
how to build a fire, how to create a stone kettle, how to cook meat.
As people moved from place to place in the landscape, performing
rituals, telling stories, and singing songs at sacred sites, they reenacted
the travels of Napi. "Significantly," wrote Gerald Oetelaar, "the total
landscape is necessary to tell the entire story, to complete the annual
ritual cycle, to establish the social and ideological continuity of the
group, and to ensure the renewal of resources."

"What you've got to realize is that the landscape is their archive,"
Oetelaar explained to me. "Those places remain alive only as long
as people visit them, remember the names, remember the stories,
remember the rituals, remember the songs."

Trail-walking cultures often grow to see the world in terms of
trails. The Western Apache believe the goal of life is to walk "the
trail of wisdom," in pursuit of three attributes Basso translated
as "smoothness of mind," "resilience of mind," and "steadiness of
mind"—puzzling phrases if viewed independently, but perfectly clear
when viewed within a metaphorical context of someone walking

---

[*] For the sake of simplicity, I have conspicuously omitted boats—both the canoes
and kayaks of Native Americans and the ships of the European colonizers—
from this explanation, and elsewhere in this book.

(smoothly, steadily, resiliently) along a trail. The Cree model for an ideal life is called the "Sweetgrass Trail," while the Navajos' ultimate good is a state of peace and balance they describe as "walking the beauty way." A creation story among the Creeks tells how their bellicose ancestors followed a "white path"—a path lined with white grass—which led them across the mountains to their current home, where they encountered a tribe of peaceful people, who were said to have white hearts. The Creeks never fully abandoned violence, but they nevertheless strove to walk the white path.

Among the Cherokee, the proper state of being for an individual is called *osi*, and the ideal state of all things is called *tohi*. According to Tom Belt, the words *osi* and *tohi* have no direct translation in English. *Osi* refers to the quality of a person who is poised on a single point of balance, centered, upright, and facing forward. *Tohi* denotes something, or everything, that is moving at its own speed, utterly at peace. An old man shuffling along the sidewalk can be *tohi*, as can a young warrior running at breakneck speed. Belt compared it to the flow of a stream, which runs fast one moment and slow the next, always moving exactly the pace that the land demands. When combined, an image of the ideal emerges: a person, upright, balanced, moving at a natural gait. Such a person is on what Cherokees call the Right Path, *du yu ko dv i*.

I asked Belt how physical trails—the dirt-and-stone paths people actually walk—figured into this metaphorical framework. Belt said that when he and his father would go out hunting in the woods, his father would always take time to describe what events had occurred in a given place. "*Someone lived here. Someone killed a deer here.* On and on and on," he said. "There's always some story about that place that's going to reconnect you with it and make it yours."

What, ultimately, connects us to land? For most animals, I suspect the answer is a mixture of mental familiarity and symbolic marking. The deer stumbles into a strange new field. It begins to explore, with halting steps, stopping often to sniff, to peer, to listen. Over time, though, it comes to recognize certain features. It learns the location of good forage. It marks certain areas with pheromones, which act as chemical signposts. It begins to move more fluidly. Lines of least resistance are discovered. Along these flow lines, with enough trips, a trail appears.

When humans make themselves at home in a new landscape, they initially behave much like deer—seeking out resources, learning routes, making signs—but over time, that field acquires an additional layer of significance. The land grows to contain not just resources, but stories, spirits, sacred nodes, and the bones of ancestors. At the same time, a deep recognition grows among the people that their lives depend on the products of the soil. People and land become interwoven, until they are nearly indistinguishable: it is no accident that, according to a striking number of cultures around the world, the first humans were sculpted from mud or clay. One version of the Hopi creation story tells that humans were created by two gods named Tawa and the Spider Woman; Tawa thought up the notion of the first man and woman, and then Spider Woman fashioned them from mud, declaring, "May the Thought live."

The trails we create from the soil are likewise born of a mixture of mud and thought. Over time, more thoughts accrete, like footprints, and new layers of significance form. Rather than mere traces of movement, trails became cultural through-lines, connecting people and places and stories—linking the trail-walker's world into a coherent, if fragile, whole.

CHAPTER 5

THE MODERN hiking trail is an uncanny thing. We hikers generally assume it is an ancient, earthborn creation—as old as dirt. But in truth, hiking was invented by nature-starved urbanites in the last three hundred years, and trails have sprouted new shapes to fulfill their hunger. To properly understand the nature of a hiking trail, one must trace the origins of that yearning, back through those early hikers to their ancestors, who set off the chain of innovations and calamities that would gradually distance humans from the planet that birthed them.

I once asked a young Cherokee woman named Yolanda Saunooke, who works at the Tribal Historic Preservation Office of the Eastern Band of Cherokee Indians, if she knew any hikers. She thought for a moment and then replied that she and her friends had spent much of their childhoods running around in the woods. "I don't know if that's considered hiking—playing on your own land, considering that it's mountainous . . ." she said. That phrase, "on your own land," snagged in the tissue of my brain. Could one go hiking on one's own land? If so, what differentiates a hike from a very long walk?

I asked some of my fellow hikers, and they all agreed that hiking on one's own land would be rather like camping in one's backyard, a kind of pantomime of the real experience. A true hike requires wilderness—land *outside* of one's (or anyone's) land.* The land must meet certain additional conditions: it must be both remote and reachable; it must be devoid of enemies or bandits, but also free of too many tourists or technology; and, most important, it must be deemed worth exploring—which is to say, people must first have learned how to derive *worth* from it, be it aesthetic or aerobic. This collision of circumstances only occurred in the modern era, when the mechanical creep of industrialism both gave us greater access to the wild and rendered it a vanishing, cherished commodity.

It is no mere coincidence, then, that the English verb *to hike*, meaning "to walk for pleasure in open country," dates back just two hundred years, nor that *hiking*, used as a gerund, only appeared in the twentieth century. Prior to that shift, the meaning of the word *hike* fell somewhere between "to sneak" and "to schlep." The com-

---

*It is important to note here that the notion of the American wilderness being unowned is also a relatively modern European belief. Of course the American wilderness once *was* owned—at least, in the usufruct sense— by Native peoples. As described in Mark David Spence's *Dispossessing the Wilderness*, prior to the Civil War, many Euro-Americans conceived of the American West as an "Indian wilderness"—a concept that was possible only because Native Americans were considered natural (i.e., not fully human) beings. However, as the conservation movement gained momentum in the late nineteenth century, the Native Americans' active management of the land (through hunting, gathering, small-scale agriculture, and the strategic use of fire) was seen as ruining the "pristine" and "primordial" qualities that conservationists had grown to cherish. William Cronon aptly captured the irony of this shift: "The myth of the wilderness as 'virgin,' uninhabited land had always been especially cruel when seen from the perspective of the Indians who had once called that land home. Now they were forced to move elsewhere, with the result that tourists could safely enjoy the illusion that they were seeing their nation in its pristine, original state, in the new morning of God's own creation."

mand to "take a hike!" (as in, "scram!") is a remnant of this older meaning. The history of how we transitioned from the one sense of the word to the other is, in some sense, the story of how modern people, and our trails, grew to finally embrace that strange thing we call wilderness.

+

In all the weeks I spent in Cherokee country, I only met one Cherokee hiker: Gilliam Jackson, the (aforementioned) administrator of the Cherokee-language Kituwah Academy. Lamar Marshall had put us in touch, saying that Jackson was renowned for "going on some of the most killer marches through the Smoky Mountains that you ever heard of." This proved to be no great exaggeration. Jackson told me he had hiked as many as forty-eight miles in a single day, and he estimated that he walked a thousand miles a year.

When I met him, he was planning to embark on a thru-hike of the Appalachian Trail to celebrate his retirement. If he was successful, he believed he would be the first full-blooded Cherokee to thru-hike the whole trail.[*]

I once asked Jackson why so few Cherokees hike. He thought this over for a moment, then replied, "I think that life has always been such a struggle on the reservation, that just survival was the biggie." Jackson had grown up in a small cabin—just "a box"—at the foot of the Snowbird Mountains, forty miles west of the reservation. His ancestors had managed to avoid the Removal by hiding in those mountains. Jackson was the third oldest of seven siblings: William,

---

[*] I checked with the Appalachian Trail Conservancy, which keeps detailed records of all the thru-hikers who registered their hikes, and they told me that out of the fourteen thousand total thru-hikers, thirteen had self-identified as American Indian, and two as Cherokee. However, there was no way to know whether those people were one-half Native American, one-quarter, one-sixty-fourth, or only Native at heart.

Lou, Shirley, Jacob, Ethel, and Esther. The whole family slept in the same room, with half the children in bed with one parent, and the other half in bed with the other. (Jackson joked that he had no idea how his parents ever found time to make more babies.) He and his siblings ran around the forest barefoot all summer. Every afternoon, it was his chore to collect firewood for the stove. His mother supplemented their meals of beans and bread with food they gathered from the woods: stewed venison or squirrel, mushrooms, and wild greens like sochan, ramps, poke, and branch lettuce. At night, they would glob pinesap onto the end of sticks and ignite them to use as torches. "Probably from the day I was born I've always been in the woods," he said.

In his teens, Jackson began exploring the trails in his area, borrowing his uncle's truck to embark on long hikes, equipped with only a wool blanket and some food pilfered from the pantry in his rucksack. He didn't remember what made him start hiking at a time when most other Cherokees didn't; he just enjoyed being out on the trail. In college, he met a group of outdoorsy white friends, and his hiking trips grew longer and longer. He eventually ran seven marathons, won a national whitewater canoeing competition, and helped found an adventure camp for at-risk Cherokee teens, which ran for twenty years before the funding dried up.

Every time I traveled back to the mountains of North Carolina, I would set aside a day to take a hike with Jackson. I loved walking with him. He hiked at a fast clip, but he paused frequently to point out plants I might have otherwise overlooked: wild iris, Indian pipe, coral mushrooms, and an odd flower called a pipsissewa, which resembled a doleful white eyeball staring at its own roots. He broke off a leaf from a sourwood tree for me to taste, and he yanked out a sassafras root, which smelled strongly of root beer. On one hike, he spotted a hen-of-the-woods mushroom, which resembled the brain of a whale: huge, gray, and labyrinthine. He carefully cut it out, took it home, soaked it in salt water to draw the insects out, and then pan-fried it in butter.

While we hiked, we often talked about the Appalachian Trail. He had countless questions for me about the logistics of a thru-hike. I warned him that opinions differed wildly, but I nevertheless had a few pieces of ironclad advice: pack light, eat healthy, and hike from south to north. (Starting out from craggy Mount Katahdin and finishing on the rolling green hills of Georgia, I opined, is like hiking from Mount Doom back to the Shire: pure anticlimax.) Finally, I advised him to arrange for some friends to meet him at various points along the trail. When you get into the middle stretches of a thru-hike, after the initial fizz has faded and before the end begins to assert its gravitational pull; when the trees leaf out and you begin to hallucinate that you are being squeezed through a giant green intestinal track; when your hips scab over and your feet swell to Flintstonian proportions; when you push hard to escape a state like Pennsylvania only to reach a state like New Jersey; when you inevitably lose sight of the purpose of the whole lunatic enterprise and just want to go home—it helps to have a few friends to cheer you along.

+

Two years after we first met, Jackson announced on his Facebook page that he would be leaving for Springer Mountain that March to begin his long-awaited thru-hike. I wrote to congratulate him, and to ask him if he would like me to accompany him for a stretch.

In June, on the day we had arranged to meet up, I stepped down off the bus in Hanover, New Hampshire, amid a cold rain. Jackson was waiting for me beneath the eaves of a nearby university building, looking like what he was: a man in his mid-sixties who had just walked seventeen hundred miles. Almost thirty pounds lighter than when I last saw him, he had grown cowl-eyed, concave in the cheeks. He was wearing mud-soaked low-top hiking shoes, synthetic cargo shorts, a zip-neck merino shirt with a hole in the elbow (chewed through, he said, by a mouse), and a battered baseball cap decorated with pins he'd collected along the trail. All of it was bubbled in a clear plastic poncho. Thrice in the next three hours, he revisited the details of how he had received a complimentary lunch at a pizzeria that day. That outsized gratitude for free food was the clearest sign of all that he had become a thru-hiker.

He retrieved his cell phone, which was charging at an electrical outlet on the wall, and I stole a garbage bag out of an empty trashcan to double-waterproof my sleeping bag, having already slipped back into the thru-hiker's habit of unapologetic scavenging. Jackson asked me what my trail name had been when I'd thru-hiked back in 2009. I told him it was "Spaceman." He said he had given himself the name "Doyi," which is the Cherokee word for "outside." The name conjured fond memories of his two-year-old grandson, Jakob, who also loved the outdoors. When Jakob was being fussy about getting dressed, all Jackson has to say was "Doyi," and the boy would come running.

From that moment on, I was Spaceman, and Jackson was Doyi.

We set out through the inundated streets. Before we even reached the trail, our shoes were squelching. Doyi told me this had been a wet year—one of those, like the year I thru-hiked, where the trail inexplicably grows sullen, and for each day spent atop sunny mountaintops, two are spent in a damp catacomb of trees. "I'm tired," Doyi confessed. "Tired of putting on wet socks, wet shoes."

After some searching, we found the trail, which led us around an athletic field and then drew us up into the darkened forest. Starting out, I was worried I wouldn't be able to keep up with his pace. I was feeling out of shape, my feet city-soft, while Doyi had been averaging twenty miles a day, an impressive pace, especially for a man his age.

However, even before the first mile had passed, it became clear that the combination of prolonged malnutrition and overexertion had sapped Doyi of his former strength; he panted on the uphills and cringed on the downhills, favoring his right knee. Walking behind him, I stared at his calves, which were hairless and lean. His body was visibly consuming itself. At one point, while climbing a moderately steep rise, he turned back to look at me and asked, between huffs, "How come you're not breathing hard?"

After an hour, a group of other thru-hikers caught up and trotted past us with the light gait of spooked deer. Each of them was white and young, with a long brown beard, thin legs, and a tidy backpack protected by a raincover: the prototypical American thru-hiker. Among their ranks, Doyi stood out; he had darker skin, had no beard, was older by decades, and was carrying a great deal more stuff.

The clouds soon ceased raining on the canopy above, but that did not stop the canopy from raining on us. It grew quiet and warm. The ground, carpeted with brown leaves and orange pine needles, released a comforting aroma. Above, thrushes sang. A barred owl said, "oo—oo—oo-ooo."

Beside the trail ran an old gray stone fence and above it lurked a pair of obese white pines, their branches weirdly splayed. Both the fence and the trees were easily overlooked, but they were clues pointing to a former era of near-total deforestation. While clearing the forest, New England farmers in the eighteenth and nineteenth centuries often spared a handful of large trees on the periphery of their property to provide shade for their livestock. As lone survivors, the spared trees luxuriated in sunlight and spread their crowns broadly. Some, like these two, were then infested with weevils, which deformed their limbs. (Trees like these were nicknamed 'wolf trees,' apparently because they greedily consumed sunlight that might benefit younger trees, as wolves devour livestock.)

Stone fences, too, were a sign that this forest had once been farmland. Large flat stones, turned up by the plow, provided a cheap, plentiful, and long-lasting building material. However, since they were painstaking to construct, stone fences only gained widespread popularity in the nineteenth century, when most of the durable hardwoods had already become prohibitively expensive. In New Hampshire—for a long time the most heavily lumbered state in the country—miles of stone fences rose as the trees fell.

Beginning in the 1920s, though, with the decline of small-scale farms and the continued rise of industrialism (and, subsequently, conservationism), much of the forest began to grow back; today, ninety percent of New Hampshire is again covered by trees. Those forests remain haunted by stone fences and wolf trees, reminders of an era when wilderness almost vanished from the region altogether.

Some hikers feel that these remnants of agriculture diminish our experience of the wild; they would prefer to walk through an old-growth forest, the older the better. But while little can surpass the grandeur and ecological complexity of a primordial forest, there is also something undeniably exhilarating about the sight of a sapling sending its shoots through the cracks in an old stone fence. It offers proof that a wild space can claw its way back against the seemingly inevitable flow of agro-industrial progress. "The creation of new wilderness in the full sense of the word is impossible," declared Aldo Leopold in 1949. But the forests of New England prove this is not always so. Walking through them—wolf trees, walls, and all—one starts to realize that the only thing more beautiful than an ancient wilderness is a new one.

+

To understand how the Appalachian Trail came to exist, it is important to know the origin of that stone fence and those deformed trees. Somewhat paradoxically, clearing fields was the first step to preserving forests. This strange transformation—from struggling to conquer the wilderness to fighting to preserve it—began before European settlers even arrived in North America.

Europeans colonized the Americas for three interlocking reasons: to send over large numbers of people, thus easing the pressures of their own overcrowded and polluted homelands; to extract and ship home previously unimagined amounts of wealth; and to tame a land they perceived as wild, wicked, and wasteful. One of their chief justifications for seizing ever-larger tracts of Native land was, somewhat ironically, that the indigenous population had failed to "improve" the land through agriculture, thereby forfeiting their rights to ownership—an argument that conveniently overlooked the fact that Native Americans had been meticulously optimizing the land to their needs for millennia.

Archaeological research suggests that the first human inhabitants of North America, including the ancestors of the Cherokee, arrived by foot, likely crossing a land bridge spanning the Bering Strait more than twenty thousand years ago. They traveled south through cool grasslands (skirting a massive glacial ice sheet, which covered most of modern-day Canada), moving from encampment to encampment, hunting gargantuan, slow-moving herbivores—mammoths, mastodons, giant bison, bear-sized beavers—with relative ease. As they moved, they learned the landscape, memorizing its plants and animals, familiarizing themselves with its weather (not just the seasonal month-to-month cycles, but also year-to-year and decade-to-decade). They likely made some irreversible changes—some archaeological evidence suggests that the Paleo-Indians were partly responsible for the extinction of many species of megafauna—but they eventually found a lifestyle that fit the contours of the land, a mixture of hunting, gathering, and (increasingly, as they moved south) farming.

The tribes of the Northeast kept their population density low and roamed widely. Their lands were open, unfenced. They burned off the forest underbrush to provide habitat to deer, elk, and bison. They planted their corn with beans and squash to shade the soil and replenish its nutrients. Land and culture intermeshed; as opposed to following a calendar filled with the names of dead emperors (Julius,

Augustus), arcane rites (Februa), and superannuated gods (Janus, Mars), they named their months after ecological cycles: the time when salmon leap upstream, when geese molt, when eggs are laid, when bears hibernate, or when corn must be planted.

In the sixteenth and seventeenth centuries, Europeans arrived on this continent bearing a radically different land ethic. The first colonists to reach America stepped down from their ships like extraterrestrials upon a so-called "new world." The god of these pale aliens had told them that life on earth was created for their use, and had instructed them to "fill the earth and subdue it." A land that was not being aggressively farmed, grazed, logged, or mined was deemed a "waste." Ownership was defined by transformation. The native people, with their shared land rights and their slow, subtle process of ecosystem engineering (too slow and subtle, it turned out, for Europeans to recognize), appeared to own no more land than any other woodland creature. "Their land is spacious and void," commented one Puritan minister, "and they are few and do but run over the grass, as do also the foxes and wild beasts."

The first English colonizers to reach this continent behaved like teenagers wandering into a quiet mansion. After searching for any plunderable goods (gold, lumber, fur), they set about rearranging the place. They cleared forests and fenced fields for English-style farming, built English-style homes, mills, and churches, and they bestowed English place names (often harking back to English places, like Hampshire). They recognized the land's bounty and grandeur, but they largely ignored the work the indigenous people had put into making it that way; having come from a place where most of the large trees had been razed, they wondered at the towering forests, without realizing they were coaxed upward by Native hands; they exulted at the profusion of wild deer without realizing that they were the result of tactical brush fires and careful hunting. Unlike the Native Americans, British farmers quickly depleted the soil; then they began hauling fish by the millions from the rivers and sea to use as fertilizer (which resulted in what one traveler called an "almost intolerable fetor"). With rifles, they hunted deer and elk almost to a vanishing point. Using iron saws, they cut down trees in enormous quantities, setting aside the larger pines to build the masts of their ships, which in turn allowed yet more people to cross the vast ocean.

Back home, England was in a dismal state. Ironically, it was the fence and the tree saw that had helped lead to this decline. During the late seventeenth century, at the same time that King Charles was selling off the royal forests to wealthy landowners, aristocrats began a process known as "enclosure"—the fencing off of once commonly held farm and grazing lands. Enclosure boosted agricultural yields but thrust thousands of peasants and laborers into a state of homelessness; in the century between 1530 and 1630, it is estimated that about half the rural peasantry were forced from their land. These new exiles flocked to cities, and then later, overseas, the ties to their ancestral land having been finally, fully severed.

At the same time, the cost of firewood in England skyrocketed, leading people to heat their homes with a cheap fuel known as sea coal. As a result, England's overcrowded cities became mired in what one writer at the time called a "Hellish and dismall Cloud." Roger Williams, the founder of what would become the state of Rhode Island, recounted that "Natives" (likely either members of the Narragansett or Wampanoag tribes) often asked him, "Why come the Englishmen hither?" The theory the Native Americans put forward was that the English had burned up all the good firewood back home, and so had crossed the ocean in search of more. They weren't far off from the truth.

The colonizers brought with them a complex form of trade we now call capitalism—the creation, exchange, and accumulation of abstract monetary value. Things and actions could be converted into money, which could be traded for other things and actions—a felled forest could become a pouch full of coins, which could become a year's supply of grain. This ingenious system allowed for near-global networks of trade. Ships were built, resources were gathered up and sold, and empires rose. People's perception of land began to subtly shift. The land was no longer merely a realm of habitation and a source of life. It was a *commodity*, whose value could and would be maximized.

The native population of North America was no stranger to trade. Long before the arrival of Europeans, the continent had been covered in vast networks of trading paths, through which flowed an enormous variety of earthly riches: salt, conch shells, feathers, flint, pigments, skins, furs, silver, copper, and pearls. With the invention of capital, though, vastly more objects could be traded in the virtual form of money—or more specifically, via strings of shells known as wampumpeag, which the colonizers popularized as a kind of universal currency. Suddenly, the continent's seemingly endless resources were opened up to the vast hunger of Europe. The foreigners' outsized craving for certain animal products—beaver pelts, deerskins, buffalo robes—drove up their price, spurring Native Americans (using European rifles and traps) to begin killing wildlife at an unprecedented rate. As the local ecosystem became less and less integral to their survival, Native peoples' impetus to carefully manage it lessened as well; if a tribe killed off all of the area's turkey or buffalo or deer, they could always buy chicken or beef from town. Some tribal communities abandoned the inland forests to reside full-time on the coast, where they stockpiled the shells necessary for making wampumpeag.

Slowly, the Native American land ethic began to fade. At the same time, many tribes were converted to Christianity, which further erased their previously reciprocal relationship to plants and animals. This cultural shift, coupled with aggressive military campaigns, dishonest treaties, and an influx of virulent diseases, resulted in what the environmental historian William Cronon described as a catastrophic mix of "economic and ecological imperialism." A vicious

cycle formed: as Native peoples' land base shrank and their traditional food sources became scarce, the pressure mounted to convert
to a more European lifestyle, which in turn consumed more land and
resources. At the same time, Europeans began to fetishize the vanishing Native population in uniquely European terms, framing them
as "noble savages," "children of Eden," and, later, as what Shepard
Krech III calls "ecological Indians": perfectly harmonious inhabitants
of the wilderness, physical and spiritual embodiments of everything
many Europeans feared they were mowing over.

As more Native people were either killed off or assimilated, English people continued flooding in. The new world grew to resemble
the old one; the countryside was increasingly covered with fenced-
off fields of English crops and pastures of English grasses (as well as
English weeds), which were populated with English cows, English
sheep, and English pigs. The lowlands filled up with farms; the forests swarmed with lumberjacks; the oceans were raked with nets.
Unprofitable spaces were converted into profitable ones: swamps
were drained, drylands irrigated, predators exterminated. Farms became plantations. Workshops became factories. Metals were mined,
oil drilled. Everywhere, wealth began to burst forth from the earth.
From this systematic reaping of natural treasures—often harvested by
slave labor—the colonies would grow into a nation of near-unrivaled
wealth, the capital of the capitalist world.

+

One place where economic value was not immediately found was in
the remote mountain ranges. In fact, the history of mountain climbing in the United States can be told as the story of people seeking
new and ever more rarified forms of value to extract from the mountains. First came the treasure-hunters, looking for precious jewels and
metals. They came home empty-handed, and the mountains were
again left alone. Next, in waves, came scientists in search of knowl-

edge, artists and writers in pursuit of beauty, tourists seeking rough pleasures, pedestrians seeking glowing health, and finally, modern outdoor enthusiasts, in pursuit of some ineffable combination of all these things.

A hundred miles north of where Doyi and I walked lay Mount Washington, the highest mountain in the Northeast, whose crest had been touched by each of these waves. White people have been climbing the "White Hill" almost as long as they have been on the continent. The first recorded ascent took place in 1642, a mere two decades after the landing at Plymouth Rock. The climb was led by an illiterate immigrant named Darby Field, whose intentions remain largely unknown. It can be assumed, however, that he did not climb it for sheer pleasure, for almost without exception, the colonists regarded mountains as either a nuisance or a horror.

Many indigenous people of the Northeast (unlike some tribes to the south and west) also avoided mountaintops, believing they were the abodes of powerful spirits. Within an animist cultural framework, this was a wholly sensible belief: What else but a mad spirit could reside in such otherworldly places? According to the toponymist Philippe Charland, long before Europeans dubbed the region's highest peak Mount Washington, the Abenaki called it *Kôdaakwajo*, or "Hidden Mountain," because its summit was so often lost in clouds. Presumably, some curious Abenakis had at some point traveled up into that misty realm and never come back down, while others had returned only to describe horrors: freak storms, shredding winds, blinding snow. (Scientists would, in fact, later record the highest wind speed on earth—excluding hurricanes or tornadoes—atop the peak.) Why would one risk going there?

The scholar Nicholas Howe has theorized that Field's true mission in climbing the peak was to show the local Abenaki Indians that white men were not subject to the same natural laws as Native Americans. The siege was, in other words, a form of psychological warfare.

With snow still on the mountaintops, Field left his home near the coast one day—accompanied by several unidentified Native guides—and followed the Saco River to the base of the White Hill. There, he discovered a village of two hundred Native Americans of an unspecified tribe. He tried to procure a mountain guide, but they refused. All but one or two of his original companions would also eventually abandon the expedition. Undaunted, Field pushed on to the summit, where, according to one account, he sat in fear for five hours, "the clouds passing under him makinge a terrible noyse against the mountains." On those cold, clear heights, he found glittering gems in the rock, which he believed to be diamonds. He returned to the summit a month later with a group of white settlers, who brought back samples of the crystals, only to discover that they were mere quartz and mica.

For the next one hundred and fifty years, no one else recorded having climbed the White Hill. In the meantime, interest in the mountains was growing among a small cadre of scientists and theologians, who regarded mountains as a potential source of new data and knowledge. The mountain was next climbed by a team of scientists in 1784, led by a clergyman-botanist named Manasseh Cutler and a clergyman-historian named Jeremy Belknap. Soon after, the peak received its presidential moniker (possibly from Belknap), and its reputation began to grow as the most "majestic" mountain in the new nation.

By the 1790s, a rough wagon road—following, as always, an old Native trail—was opened through a notch in the White Mountain chain along the western flank of Mount Washington. As it was gradually widened and improved, the road provided the most direct route from southern New England to northwest New Hampshire and Maine—"a great artery," wrote Nathaniel Hawthorne, "through which the life-blood of internal commerce continually throbbed."

Around the turn of the nineteenth century, a young woodsman

named Abel Crawford decided to open an inn alongside that road and began guiding curious adventurers up the mountain, where they could enjoy a sublime panorama of the surrounding mountains. To facilitate the trip to the summit, Abel and his son Ethan cut trails. The first of these, called the Crawford Path, might well be the oldest continuously used hiking trail in the country. It is a slow, circuitous path, winding slowly back and forth up across the mountain "as if reluctant to approach too directly into such an august presence," wrote Laura and Guy Waterman in *Forest and Crag*, their author-itative history of hiking in the Northeast.* At first, the path was faint—one early hiker described it as "obscure, often determined only by marked trees, some of which 'Old Crawford' alone could discover"—but over time it became clear and wide. More than a century later, the last leg of this path would become part of the Appalachian Trail.

A new curiosity and admiration for the mountains was taking hold, and Mount Washington loomed prominently. The peak was climbed by Emerson, Hawthorne, and Thoreau (twice). They all seemed to recognize some twinkling of the divine in it. Hawthorne found it "majestic, and even awful"—not horrible, but full of awe. Thoreau, writing to a friend who had recently climbed the peak, wrote, "You must have been enriched by your solitary walk over the mountains. I suppose that I feel the same awe when on their summits that many do on entering a church."

By the 1830s, the barren mountains were being valued for the same reasons they had once been reviled: their terrific heights, their unpredictable weather, and perhaps most of all, their remoteness from the lowland clutter of civilization. Like storm clouds, slowly, and

---

* The Watermans' statement was somewhat coy. Abel and Ethan most likely cut the path in a gradual fashion because they knew that it would one day need to suit both skittish horses and wilting urbanites.

then all at once, around the mountaintop an aesthetic appreciation had coalesced. "It became almost an obligatory mark of a vigorous public man of New England in those years that he had made the ascent of Mount Washington," wrote the Watermans. The sentiment seems to have originated among city dwellers, for whom mountains were exotic. The people who lived at the base of the peaks—who were necessarily fixated on extracting economic and subsistence value from the land—were unlikely to ever climb them. One farmer at the base of Mount Washington told the pastor Thomas Starr King that he wished the mountains were flat.

Many urban tourists were eager to see the mountains, but were unable or unwilling to walk up them. So, in 1840, the Crawfords widened their path to make it fit for horses; Abel, then seventy-four, was the first man to ride on horseback all the way to the summit. By the 1850s, all five paths up Mount Washington had been converted into horse paths. A decade later, a carriage path had been cleared, and around the same time, another path cut by the Crawfords was used to lay out the tracks of a cog railway. It became possible to travel from the back alleys of Boston to the top of Mount Washington without taking more than a few steps. One prominent writer recommended that his readers catch a train to Gorham, take a wagon "through primeval forests" to the Glen House Hotel, and then ascend the mountain on a pony. "That is *de rigeur*," he insisted. With newfound ease and expediency, as many as five thousand people reached the summit of Mount Washington each year.

In the early days, if one of their clients had wanted to spend the night on the mountainside, the Crawfords would have built a pole-and-bark shelter for them, and inside, they would have fashioned a bed of fragrant balsam boughs. By the 1850s, to accommodate the new flood of tourists, two stone-walled hotels—aptly named the Tip-Top House and the Summit House—were built directly on top of the peak. On mountaintops throughout the Northeast, similar buildings—hotels, huts, concession stands, even a small newspaper office—were popping up like mushrooms. Meanwhile, vast vacation resorts sprawled across the valleys; one hotel, built in the Catskills, boasted a thousand rooms. Guests were known to spend whole summers at these "grand hotels," taking short day trips out into the mountains to amuse themselves. Walking paths tendriled out around the hotels, many of them equipped with wooden ladders, scenic overlooks, and designated resting areas.

The Civil War brought a decades-long drought to mountain tourism. But around the turn of the century, the arrival of the automobile granted people easy access to previously unreachable mountains, and the public's interest in hiking revived. A slew of hiking clubs formed to maintain old trails and build new ones. Meanwhile, for those disinclined to walk, a series of road improvements made it possible to drive right to the crest of Mount Washington. Like anywhere cars and tourists converge, a large souvenir shop and a cafeteria opened up to cater to the crowds. To this day, atop that storied peak, drivers can be found proudly purchasing bumper stickers that read: "This Car Climbed Mt. Washington."

When I reached the summit of Mount Washington on my own thru-hike, it struck me as a kind of suburban horror. As I neared the summit, a red-and-white radio antenna rose into view, followed in time by a stone tower, a cog railroad, a cafeteria, and a crowded parking lot. It was a clear, warm Saturday in July, and the peak writhed with tourists. After four months of walking over more or less barren peaks, it felt like I had stumbled into an outdoor mall.

Almost four hundred years earlier, Darby Field had deemed Mount Washington a barren wasteland, devoid of economic value. In the intervening centuries, the peak's barrenness had served as a beacon for hikers, offering a rare island of wilderness in a sea of tamed fields. As one hiker wrote in 1882, "The climber here tastes the full enjoyment of an encounter with untamed nature." It was a perverse fate, then, that the mountain's untamed allure would be precisely what led to its own taming.

Little did I know that, but for a few flukes of history and a widespread shift in popular sentiment, many of the other peaks I had crossed on the Appalachian Trail could have looked the same.

+

A few hundred miles north of Mount Washington stands its wild twin, Mount Katahdin. At the outset of the American experiment, the two were not dissimilar. Like Washington, indigenous people living near the base of Katahdin reportedly never climbed it, for fear of a winged thunder spirit named Pamola. Both mountains would one day be recognized as the tallest in their respective states: indeed, the Penobscot name for Katahdin means "the greatest mountain." From similar beginnings, though, the two parted ways soon after colonists arrived. While the slopes of Mount Washington experienced ever-growing waves of visitors, Katahdin, walled off by miles of the gloomy North Woods, remained unclimbed. It was finally crested by an eleven-man team of government surveyors in 1804, more than a century and a half after Mount Washington.

In 1846, Thoreau made a failed bid to climb Katahdin. He and two companions made their way to its base by canoe, guided by an old Native American man named Louis Neptune, who advised Thoreau to leave a bottle of rum on top of the mountain to appease the mountain spirit. On their climb, Thoreau and his companions followed moose trails and scrambled cross-country. In one harrowing instance, while crawling over the flattened tops of the black spruce trees that had grown up between the mountain's massive boulders, Thoreau looked down to find that below him, in the crevices, lay the sleeping forms of bears. ("Certainly the most treacherous and porous country I ever traveled," he wryly observed.)

The party became lost in fog and never made it to the summit.

But on his descent, passing through an area called the Burnt Lands, Thoreau—who had spent almost his entire life in bucolic Concord, where a surplus of farms and fences had rendered the landscape "tame and cheap"—suddenly realized he had stumbled upon a wholly wild place. He found the Burnt Lands savage, awful, and unspeakably beautiful. Here, he sensed, was the universal bedrock underlying the artifices of humankind. Recalling the experience, he wrote:

> This was that Earth of which we have heard, made out of Chaos and Old Night. Here was no man's garden, but the un-handseled globe. It was not lawn, nor pasture, nor mead, nor woodland, nor lea, nor arable, nor wasteland . . . Man was not to be associated with it. It was Matter, vast, terrific . . . rocks, trees, wind on our cheeks! the *solid* earth! the *actual* world! the *common sense! Contact! Contact!*

How, one must wonder, had a human being—indeed, a whole generation of human beings—become so abstracted from the land (the *solid* earth! the *actual* world!) as to warrant such an epiphany? The answer, as we've seen, stretches back through our ancestral past: through agriculture, which obviated the hunter-gatherer's need to walk, study, and interact with whole ecosystems; through writing, which replaced the landscape as an archive of communal knowl-edge; through monotheism, which vanquished the animist spirits and erased their earthly shrines; through urbanization, which con-centrated people in built environments; and through a snug pairing of mechanical technology and animal husbandry, which allowed people to travel over the earth at blurring speeds. Euro-Americans had been working for millennia to forget what an unpeopled planet looked like. To see it afresh came as a shock.

Ever since Thoreau's revelation, a steady trickle of hikers has flowed toward Katahdin in search of the same ineffable experience.

It gained a reputation as the antithesis to mountains like Washington, where, according to one account, "large flocks of hitherto 'un-mountain-fähig', both male and female, streamed up the mountains like a transplanted tea party." But despite its growing popularity, Katahdin resisted all attempts to tame it. During the height of the summit house craze of the 1850s, Maine politicians, envious of the commercial success of Mount Washington, chartered a road to be built over Katahdin. A crew was sent out to survey a path, but they returned with a route so absurdly steep that no carriage could climb it, and the project was soon abandoned. Even into the 1890s, while trail-builders on Mount Washington were rearranging boulders to construct paths so smooth they reportedly could be walked blindfolded, the paths on Katahdin remained, in the Watermans' words, "the roughest of cuts through the north woods."

The longer Katahdin resisted attempts to tame it, the Watermans wrote, the more it attracted "pilgrims" who enjoyed its wild character—and who, moreover, would fight to keep it that way. In 1920, an eccentric millionaire named Percival Baxter climbed Katahdin via the vertiginous Knife's Edge route. Greatly impressed, he vowed to ensure that the land would remain "forever wild." The following year, as governor of the state, he fought to have the area recognized as a state park. When the state legislature refused, he began buying the land with his own fortune, eventually acquiring two hundred thousand acres, which was later designated as a state park. From the outset, Baxter insisted that "Everything in connection with the Park must be left simple and natural and must remain as nearly as possible as it was when only the Indians and the animals roamed at will through these areas."

Ten years later, there was a push by Baxter's political nemesis, Owen Brewster, to make the area look more like the White Mountains, by building new motor roads (now feasible, thanks to technological advances) along with a large lodge and a series of smaller cabins. Baxter successfully fought them back, and the park remained stubbornly inaccessible. The park's wildness, in other words, was not given. It was made.

It may sound strange (even sacrilegious) to some, but in a very real way, wilderness is a human creation. We create it in the same sense that we create trails; we do not create the soil or the plants, the geology or the topology (although we can, and do, shift these things). Instead, we delineate the place, by defining its boundaries, its meaning, and its use. The history of Katahdin is emblematic of the wilderness as a whole, which has always been the direct result of human ingenuity, foresight, and restraint.

"Civilization," wrote the historian Roderick Nash, "invented wilderness." According to his account, the wilderness was born at the dawn of agro-pastoralism, when we began cleaving the world into the binary categories of wild and tame, natural and cultivated. Words for wilderness are notably absent among the languages of hunter-gatherer peoples. ("Only to the white man," wrote Luther Standing Bear, "was nature a wilderness.") From the vantage point of a farmer, the wilderness was a strange, barren land, full of poisonous plants and deadly animals, antithetical to the warmth and security of home. To these land-tamers, wilderness became synonymous with confusion, wickedness, and suffering. William Bradford, the governor of the Plymouth Colony, was representative of this mindset when he deemed the uncolonized countryside "a hideous and desolate wilderness full of wild beasts and wild men."

For centuries after the rise of agriculture, we erected fences to keep our cultivated land safe from whatever lurked in the darkness. But the realm of cultivation continued to spread, insatiably, until it at last began to endanger the wilderness, rather than the other way around. Then we began fencing *in* the wild, to keep it safe from us. For obvious reasons, this shift came much earlier on the isle of Britain—which began walling off its forests a thousand years ago—than it did on the seemingly endless American continent.

Amid the coal-fired fug of industrialism, people began to recognize that the unchecked spread of civilization could be toxic, and the wilderness, by comparison, came to represent cleanliness and health. Quite suddenly, the symbolic polarity of the word *wilderness* was reversed: it went from being wicked to being holy. That switch allowed a new set of moral attitudes toward the non-human world to take hold. Even a man as wilderness-averse as Aldous Huxley came to understand that "a man misses something by not establishing a participative and living relationship with the non-human world of animals and plants, landscapes and stars and seasons. By failing to be, vicariously, the not-self, he fails to be completely himself."

This is the most succinct definition of the wilderness I have found: the *not*-self. There, in the one place we have not remolded in our image, a very deep and ancient form of wisdom can be found. "At the heart of all beauty lies something inhuman," wrote Albert Camus. We glimpse this inhuman heart only once the rosy lens of familiarity has fallen away. Then, Camus wrote, we realize that the world is "foreign and irreducible to us"—a sensation acutely familiar to both Thoreau and Huxley. "These hills, the softness of the sky, the outline of these trees at this very minute lose the illusory meaning with which we had clothed them," he wrote. "The primitive hostility of the world rises up to face us across millennia." We over-civilized humans cherish wilderness because it both fosters and embodies that sense of not-self—it is a brazenly naked land, where a person, in mingled fear and awe, verging on nonsense, can cry out: *Contact!*

+

Doyi and I followed the Appalachian Trail northward. We climbed up and over a bulge called Moose Mountain, falling into an easy rhythm. The trail bore a string of deep moose prints and a pile of olivey pellets, but no moose. The view from the summit was the same as from the window of a cloud-socked airplane. On the downhill side, the wind shouldered through the trees, shaking down leaves and water. We were glad to reach the lean-to—a wooden shelter, ubiquitous on the trail, shaped like a heavily italicized letter L.* Someone had strung up a tarp over the entrance to keep out the wind and rain.

"Hello? Anybody in there?" Doyi called.

"Doyi!" voices cheered, in unison.

Inside it was dark, steamy, sour-smelling. The thru-hikers were all burrowed in their sleeping bags, some leaned upright against the back wall, others supine. Headlamps blazed coldly from the center of their foreheads. Doyi introduced me to them, from right to left: Gingko, an albinic young German man with an ice-white beard and startling blue eyes; Socks, a cheerful, dark-haired young woman, so named after her resemblance to Sacajawea, though she was Korean-American; Catch-Me-If-You-Can, a Korean-American man in his forties, quiet, high-cheek-boned, forever smiling, and renowned for his speed; and Tree Frog, a young white man with bushy brown hair, who often told strangers along the trail he was employed as a butler, because he had learned it was more interesting to lie about being a butler than to tell them the truth about being an engineer. Doyi had known some of them for months and others only a few days, but he had an easy rapport with all of them. As we dropped our packs inside the shelter, he asked them to please scooch over and make room for us. Distracted, they were slow in moving, so he joked, "Don't worry, you don't have to do it right away. Anytime in the next ten seconds would be fine." They laughed, and then moved over.

Doyi and I changed clothes, got in our sleeping bags, and prepared dinner. Tree Frog said that as he hiked he had been practicing

---

* The shape of these structures, properly called Adirondack lean-tos, was inspired by the kinds of impromptu bark shelters that mountain guides like the Crawfords once built for their clients.

the Cherokee words that Doyi had taught him: "shit" (*di ga si*), "shit!" (*e ha*), "water" (*ama*), and *Osda Nigada*, which means, roughly, "It's all good." *Osda Nigada!* had become a kind of rallying cry for the rain-drenched hikers, and soon became their unofficial name for themselves: Team Osda Nigada.

As I sat over my Coke-can stove cooking a pot of soba noodles, I found myself slipping back into the headspace of a thru-hiker. Tree Frog generously offered me and Doyi two muffins he'd carried up from town. They were sticky and dense; we both scraped the muffin paper clean with our teeth. (Nothing tastes better, the old thru-hiker adage says, than food you haven't had to carry.) The gift prompted Doyi to teach the group a new Cherokee phrase: "*Gv Ge Yu A*," which means "I love you," except, Doyi said, that it cannot be used casually; it can only be spoken when one truly means it.

Tree Frog was bent over his journal, scribbling down the day's events. He was working on a book about his mother's attempt to thru-hike the AT, which was halted by cancer, and his subsequent quest to scatter her ashes atop Mount Katahdin. In exchange for the muffin, I offered him my dimpled copy of the latest *New Yorker*, the fiction issue. He politely waved me off. "Sounds heavy," he said. He meant the weight of the paper, not the subject matter.

They talked primarily about time and food; when they would reach certain mountains or towns or states; what they were eating, had eaten, would eat, would *like* to eat. The interior life of a thru-hiker this far into a long hike is a mixture of waning adventure-lust, intensifying hunger, mild impatience, and calm, single-pointed focus. The pull of Katahdin drew them inexorably along the same trail, at roughly the same pace, like marbles in a downward groove. They had recently agreed to try to summit Katahdin as a group, even if that meant slowing down to accommodate the slower members. After consulting his guidebook that night, Tree Frog suggested that they should try to finish by July 7. Doyi smiled at the thought of that

golden, mirrored numeral—7/7—a sacred number for the Cherokees. It had the glow of fate.

+

What makes a trail wild? Is it the people who built it, the people
who walk it, or the land around it? The answer is a combination of
all three. In large part, the Appalachian Trail gained its wild reputation from the iconic wildernesses it managed to string together:
not just Katahdin, but also the Great Smokies, the Blue Ridge, the
Cumberlands, the Greens, the Whites, the Bigelows, the 100 Mile
Wilderness. Thanks to a massive land acquisition project led by the
Appalachian Trail Conservancy, the gaps between those wilderness
areas were later filled in. Today, the trail is surrounded by an almost
uninterrupted, thousand-foot-wide belt of protected land—what is
sometimes referred to as "the longest, skinniest part of America's
national park system."

Those lands, though, would never have been protected if like-
minded hikers and activists hadn't fought for their protection. The
AT—like any trail—is the creation of multitudes: walkers, trail-
builders, conservationists, administrators, donors, and government
officials. Before all of them, however, the trail was born from the
imagination of a single man—a forester, wilderness advocate, and
utopian dreamer named Benton MacKaye. Even today, the trail bears
the imprint of his brilliant and idiosyncratic mind.

The idea for the Appalachian Trail reportedly first occurred to
MacKaye while hiking through the Green Mountains of Vermont in
1900, at the age of twenty-one. He and a friend had climbed a tree
atop Stratton Mountain to admire the view, and, dizzy with a "planetary feeling," as he later described it, MacKaye suddenly envisioned
a single trail stringing together the entire Appalachian range from
north to south. Two years later, while working at a summer camp in

New Hampshire, he mentioned the idea to his boss, who replied that it sounded like "a damn fool scheme."

History would prove otherwise. In fact, at that precise moment, disparate forces were aligning to allow something as audacious as a two-thousand-mile-long hiking trail to one day exist. In newspapers and books from the turn of the century, America was increasingly being seen as a land of worsening health, degenerating morals, and rampant money grubbing. Boys were growing too weak, while girls were "overheated, overdressed, and over-entertained." These fears stemmed in part from a rapid and unprecedented surge in urbanization; Manhattan, for instance, housed more people in 1900 than it does today. Time spent outdoors, in the "fresh air"—a newly popular phrase—was seen as a curative for society's ills. Locomotives (and soon, automobiles) made trips to the mountains easier and faster. Summer camps sprang up throughout the Northeast. (The summer camp I attended, Pine Island, was founded in 1902.) The turn of the century also marked the birth of the scouting movement. In 1902, a nature writer named Ernest Thompson Seton founded a club for boys called the League of the Woodcraft Indians, which later inspired Robert Baden-Powell to form the Boy and Girl Scouts. "This is a time," wrote Seton in 1907, "when the whole nation is turning toward the outdoor life."

Meanwhile, the federal government—at the urging of hiking-cum-conservation groups like the Appalachian Mountain Club and the Sierra Club—had begun setting aside huge tracts of public land. This process began in 1864, when Abraham Lincoln, following the advice of Frederick Law Olmsted, signed a bill setting aside the Yosemite valley and a nearby grove of giant sequoia trees as public land. Olmsted, the famed designer of Central Park—which he insisted remain open to all, "the poor and the rich, the young and the old, the vicious and the virtuous"—warned Lincoln that, if the Yosemite valley fell into private hands, it could end up as a walled garden for the sole enjoyment of the rich, like many parks in England. By signing the Yosemite Grant Act, Lincoln set a key precedent for the creation of the national park system. The conservationist movement began to reach a new peak in 1901, when Theodore Roosevelt, a lifelong outdoorsman, assumed the presidency. His first address to Congress called for the creation of a series of national forests. By the end of his presidency in 1909, he would set aside one hundred fifty national forests, fifty-one federal bird reserves, and five national parks. All told, he protected roughly 230 million acres of public land.

Meanwhile, a new conception of trails was spreading. Trail designers began to reconsider the isolated clusters of trails that had once surrounded the most popular hiking destinations and discovered ways to connect those clusters into cohesive networks. Soon, there arose the notion of a "through trail"—a trail that would keep going. In 1910, James P. Taylor, a schoolmaster who enjoyed taking his students on long hikes, proposed the construction of a single trail connecting all of the tallest mountains in Vermont. He called it "The Long Trail."

Into this intellectual environment stepped MacKaye. He graduated from Harvard a few months before Roosevelt's inauguration, and shortly after earned his master's degree from the Harvard School of Forestry. In the following decades, he took a series of forestry and planning jobs, which gave him a better sense of how people can transform landscapes (and vice versa). During one such project, in 1912, he conducted an influential study on the effects of rainwater runoff in the White Mountains, which proved that deforestation contributes to flooding. Partly as a result of his study, the White Mountains were later designated a national forest.

Over the course of twenty years, MacKaye grew from a gangly young forestry student into a bespectacled, dark-haired, hawk-faced intellectual, with a pipe permanently clenched between his teeth. All the while, his idea for what he called "an Appalachian Trail" grew along with him. In 1921, he lost his wife, Betty—a suffragist and peace advocate—when she drowned herself in Manhattan's East River. Grieving, MacKaye holed up in a friend's farmhouse in New Jersey, where he paused his forestry work long enough to put his idea for the Appalachian Trail down on paper. What emerged was more than a mere trail. The innocuous title he gave to his now-historic proposal—"An Appalachian Trail: A Project in Regional Planning"—belied its radical vision. In fact, he saw the trail as nothing less than a remedy to the worst ills of urbanization, capitalism, militarism, and industrialism—what he called "the problem of living."

MacKaye's thoughts on how to transform our society were strongly shaped by a five-hundred-page philosophical treatise called *The Economy of Happiness*, authored by his brother James. Drawing on the works of Bentham, Malthus, Darwin, Spencer, and Marx, James MacKaye sought to devise a rigorous response to the ugliest aspects of industrialism. Rather than a society of independent actors each seeking to maximize profit—which unintentionally resulted in a "vast and increasing surplus of misery"—he envisioned a steady-state economy managed by a technocratic elite, who strove to maximize the "output of happiness." Anticipating the inevitable question of how a government could possibly measure a nation's happiness, the book was littered with equations and graphs attempting to quantify well-being. (The biographer Larry Anderson quipped that it was, ironically, "possibly the most humorless and austere tract ever devoted to the subject of happiness.")

Most significantly, MacKaye's brother taught him that the key to solving societal problems was to change systems, not human nature. As MacKaye became an increasingly prominent voice in the conservation movement, he seldom wrote about greed or excess. He chose instead to focus on environments—how they can weaken us, or how they can be altered to strengthen us. Having spent much of his childhood in New York City (which he loathed), he chose to attack the ills of modernity through its most obvious manifestation: the de-natured, overpopulated, hyper-competitive metropolis.

From the outset, the overarching goal of Benton's work was to circumvent the sense of alienation that had been growing among Euro-Americans for centuries. The crucial first step, he concluded, was to secure a space outside the reach of the metropolis—"a sanctuary and a refuge from the scramble of every-day worldly commercial life"—in which people could learn to live anew. He applauded the rise of the national parks, but lamented the fact that they were all so far away; at the time, of the seventeen national parks, only one was east of the Mississippi River. Meanwhile, he wrote, a continuous green belt of wild land, the Appalachian range, lay "within a day's ride from centers containing more than half the population of the United States."

Along the trail, MacKaye wanted to build not just a string of rustic shelters, but also nonprofit wilderness camps, collective farms, and health retreats where the citizens of America's industrial centers could escape for fresh air.* The source of modern malaise, he believed, was that civilized people were no longer equipped to survive in nature. They had forgotten how to raise food, how to build things, how to travel on foot. They were entirely dependent on the economy for their survival, which led them to be overworked and unhappy. People needed to get "back to the land," MacKaye wrote.

---

* It is surely no coincidence that MacKaye, having just lost his wife to what was then called "nervous depression," stressed the importance of wilderness in maintaining mental health. A note of mad hope can be detected as he writes, "Most sanitariums now established are perfectly useless to those afflicted with mental disease—the most terrible, usually, of any disease. Many of these sufferers could be cured. But not merely by 'treatment.' They need acres not medicine. Thousands of acres of this mountain land should be devoted to them with whole communities planned and equipped for their cure."

Some elements of the proposal eventually proved surprisingly prescient. As he had envisioned, a series of rustic shelters were built along its full length, each no more than a day's hike apart. He insisted that the trail should be maintained by volunteers, not paid workers, because for volunteers "'work' is really 'play.'" And, as he astutely argued, constructing a two-thousand-mile trail was less daunting than one might think, because it need not be constructed ex nihilo. Instead, trail-builders could simply stitch together a string of existing trails, including the Long Trail, one hundred fifty miles of which would later be folded into the Appalachian Trail.

In 1927, MacKaye was invited to articulate his vision to the New England Trail Conference. The paper he delivered, entitled "Outdoor Culture: The Philosophy of Through Trails," was not quite what they had anticipated. In fiery tones, MacKaye laid out the full breadth of his plan for a connected corridor of wilderness work camps. Drawing from the example of ancient Rome, his dialectic positioned the decadent metropolis against the barbarian hinterlands. He railed against the "lollipopedness" of jazz-loving, picnic-eating city dwellers, and he contrasted these human "jellyfish" with the strong, tough, wilderness-savvy proletariat his trail would attract.

"And now I come straight to the point of the philosophy of through trails," MacKaye concluded. "*It is to organize a Barbarian invasion. It is a counter movement to the Metropolitan invasion . . . As the Civilizees are working outward from the urban centers, we Barbarians must be working downward from the mountain tops.*"

In the end, the genteel East Coast trail-building community blanched at the more utopian elements of MacKaye's vision. But work on the trail itself began in earnest. The task of actually constructing the trail, which MacKaye showed little interest in, fell largely on the shoulders of a Maine native named Myron Avery, a husky, weather-beaten pragmatist with the bearing of a football halfback. Under his leadership, the trail was completed in 1937, by linking together a chain of logging roads, old hiking paths, and hundreds of miles of fresh-cut trail. But the bulk of MacKaye's vision had been pared away. Gone were the camps, the farms, and the sanitariums. The many-limbed idea streamlined, until it emerged as a single, sinuous trail through the woods.

MacKaye eventually grew to accept the trail's new, narrower mission: to provide a "path of endless expeditions" through the wilderness. By 1971, when an interviewer asked him to state the Appalachian Trail's "ultimate purpose," MacKaye, then ninety-two and nearly blind, had whittled down his answer to Zen simplicity:

1. to walk;
2. to see; and
3. to *see* what you see!

Nevertheless, intentions have echoes. The trail's radical origins began to manifest themselves in unforeseen ways in the decades that followed, most notably in the community of hikers who, in ever-increasing waves since the end of the Second World War, undertook pilgrimages from one end to the other in search of their own answers to the problem of living. Nomadic, hirsute, and reeking, they were, and remain, the very image of MacKaye's barbarians. Come July, one can spot them lining a highway in southern New Hampshire, thumbing rides in the rain; roaming like wolves through the mammoth, icily lit grocery stores of Virginia; and shacking up, three to a bed, in a motel in Pennsylvania. Once in a while, one might even catch them in Times Square, having ridden the afternoon train in from Bear Mountain, looking at once shell-shocked and childishly delighted at the flood of light and sound. As one former thru-hiker told me, "Most people live in civilization and visit the woods. But when you're thru-hiking, you're living out in the woods and visiting civilization."

+

Snug and dry in the lean-to, Team Osda Nigada nestled down to sleep. I plugged my ears with wax to keep out the sound of snoring and the pock of the rain on the tin roof. Around ten P.M., long after sunset, the bright star of a headlamp appeared inside the lean-to. It hovered insistently above me. I unplugged my ears. "Hey," a voice said. "Sorry. Can you please move over? I don't have a tent." Grumpily, we rearranged our things to accommodate the dripping newcomer, until we were pressed shoulder-to-shoulder.

Just after sunrise, people began rustling around. Nothing had dried out overnight, despite being hung up on a clothesline. When the thru-hikers wrung out their wool socks, they produced something resembling milk coffee. Nobody took the time to cook breakfast; an energy bar, a few handfuls of trail mix, or a heaping scoop of peanut butter sufficed.

The daylight revealed the late-night arrival to be a south bounder (or a SoBo, in trail parlance)—one who was hiking south from Katahdin to Springer Mountain. In the Northern states, the SoBos were easy to spot, since unlike NoBos, they hadn't had time to grow a long beard, and because they tended to be loners. This one was no exception. He told us he had started only thirteen days ago from Katahdin. (The night before, Tree Frog had calculated that it would take them at least twenty-four days to reach Katahdin.)

Doyi did some quick mental math. "You hiked four hundred thirty miles in thirteen days?"

"Yep," was all the SoBo said, before he lightly lifted his backpack and stepped out of the shelter.

The other thru-hikers were quiet for a little while.

"He's *scootin'*," Doyi said.

"Doing thirty-mile days through Maine and New Hampshire?" Tree Frog said. "Wow."

One by one, the thru-hikers put on their wet boots and, with a sharp breath, as if plunging into cold water, stepped out of the shelter and into the clouds. I was the last one to depart. There was no sun. Plants drooped, as if hungover from the night before; a pink orchid wept.

Eager to catch up, I raced over the mountain and down the other side. At the bottom I crossed a road and entered a field of high grass, where I was startled to find a pink plastic flamingo and a handmade sign depicting a cheerful old man holding a pink ice-cream cone. The sign read: "BILL ACKERLY / HIS ICECREAM BRINGS ALL THE HIKERS TO THE YARD / HIS WATER TASTES BETTER THAN YOURS / DAMN RIGHT, HIS CROQUET GAME IS BETTER THAN YOURS / IT'S ALL FREE, YEAH THERE IS NO CHARGE!!" I followed a little side trail to find a blue house with white trim, festooned with Tibetan prayer flags. In the backyard, a pristine croquet court had been hacked out of the high grass. Doyi sat on the porch, talking with Ackerly, who got up to shake my hand. Ackerly had a long face topped with vanishing gray hair, large glasses, and a moony smile.

He asked my name. I told him.

"Spaceman?" he said, dreamily. "Like all of this beautiful space..."

We sat for a long while on Ackerly's porch, talking about Tibet (where he had visited), the works of Homer (which he had studied), and other thru-hikers (whom he had been feeding, for free, every summer for over a decade—a practice hikers call "trail magic").

Somehow, our conversation turned to Doyi and his Cherokee heritage.

"We need to honor this man here," Ackerly said, gesturing to Doyi. "He is our ancestor. His people were here first. You know, people always say Christopher Columbus was here first, but he wasn't."

"He was *lost*," Doyi said.

"That's right. Columbus was a terrible man."

Doyi nodded, gravely.

"Well, anyway," Ackerly added, "in the grand scheme of things, we're all children."

As we hoisted our packs and prepared to leave, Ackerly gave each of us a hug. Back on the trail, I asked Doyi if he found it odd that Ackerly had referred to him as "our ancestor." He brushed it off. "There are a lot of good people on this earth," he said. "What I've enjoyed most about this hike is meeting people like Bill." He was continually awestruck by the goodness the trail brought out in people. One day, when his knee was really hurting, a fellow thru-hiker had offered to carry his pack for him. Doyi passed on the offer, but he was moved nonetheless—a stranger was willing to practically double his own suffering to alleviate Doyi's. "That's the real trail magic, to me," he said. "People helping people."

+

A trail that is never used fades from existence. But in the postwar era, as hiking became increasingly popular among Americans, a new danger emerged: trails were suddenly in danger of *over*-use. By the 1970s, it was often said they were being "loved to death." Unprecedented numbers of hikers were storming the mountains wearing heavy lug-soled boots nicknamed "waffle stompers," churning up soil, which then eroded or turned to mud. The most popular trails suffered the worst damage, since more than half of all hikers used only ten percent of the trails. In the Smokies, where trails were also open to horse traffic, some of the trails were worn down chest-deep, whereas up north, where the soil is rockier, others widened to forty feet.

In response, trail-builders had started designing so-called sustainable trails, which carefully minimized erosion, avoided sensitive plant life, and prevented the contamination of nearby water sources. By the 1990s, modern hiking trails—which had already contorted themselves to reach places no other trail in history would have previously bothered going—began to take on a whole new shape and internal logic. They could no longer simply focus on reaching wild spaces; they now needed to ensure that they didn't snuff out that wild quality for future walkers.

Managing people and managing water, it turns out, are the twin challenges of designing a sustainable trail. Unfortunately for trail-builders, those two needs are not always aligned. For example, trail-builders like to install stone steps leading up steep hillsides, because steps provide a durable walking surface for hikers and break up the flow of water, which slows erosion. However, hikers tend to dislike steps, because they look unnatural and often require more work to climb. So hikers will often try to climb up the hillside bordering the staircases, which gutters rainwater and worsens erosion. This forces trail-builders to install menacingly jagged rocks, called gargoyles, on either side of the staircase.

Something similar happens on switchbacks, the long curvy turns that trail-builders create to lessen the trail's incline and slow erosion. If hikers can see from one turn to the next, they will almost inevitably create a shortcut. Among trail-builders, it is axiomatic that when hikers get tired, hikers get selfish. Many trail-builders find this tendency immensely frustrating. "I always say that this whole 'hiker management' thing would be a lot easier if we just got rid of the hikers," joked Morgan Sommerville, a former trail crew leader.

When a shortcut forms, the trail-builder's first impulse is typically to simply block it off, but that doesn't always work. In this regard, hikers behave remarkably like water; eventually, they will drip through almost any obstacle to follow the line of least resistance. Recently, Sommerville told me, to deter hikers from taking an old, degraded fall-line trail up to a mountain bald called Max Patch, a team of trail-builders had installed a large sign in the middle of the old trail pointing hikers in the direction of the new one. For good measure, on either side of the sign, they planted a row of rhododendrons. "That lasted about two or three months," Sommerville chuckled. "People just picked the most vulnerable-looking rhododendron, eliminated it, and kept going up there. I went there in October, and there was just a constant line of people walking straight up the hill. At that point, the signs asking people not to go that way had *also* been removed by . . . whomever."

A professional trail-builder named Todd Branham once told me that he too would resort to dropping a pile of branches or a big rock in a place where people would be tempted to create a shortcut. "But if the trail is well-designed," he said, "I won't have to do that, because people will *want* to stay on the trail. They'll be having so much dang fun they won't want to get *off* the trail."

The central task of the trail-builder is to navigate an age-old dilemma: to convince people to do what they *should* do (to best serve the long-term collective good), rather than what their basest instincts tell them to do (to best serve themselves in the short run). As I learned while shepherding, the easiest way to bend a group's trajectory is to accommodate their desires. That was Branham's credo, too. For example, if people can hear a waterfall but don't have a trail leading to it, he said, they will just create their own crude trail. Impromptu trails like these are notoriously difficult to get rid of, because other hikers will inevitably be drawn to them. Instead, a smart trail-builder will aim to find the most sustainable route to that waterfall in advance and then guide hikers there. That way, the trail can both preserve the integrity of the land and fulfill the hikers' desires.

To know all the potential routes a trail could take, a trail-builder must have a wide-ranging knowledge of the surrounding landscape. The first step to building a sustainable trail is to study a map of the region and gather a rough idea of where the best route might lie. Next, the trail-builder scouts out the route on foot. I watched Branham one day while he went through this process in a patch of woods in Brevard, North Carolina. He began by walking the proposed route of the trail, feeling out the angle of ascent and the quality of the soil. He tried multiple iterations of each line, searching for the most graceful approach. (Watching him pace up and down the hillside, I was reminded of how ants will try out multiple routes before settling on the best one. He was doing the same thing, only far in advance.) Then, using a roll of orange plastic tape, he "flagged" each section of the trail by tying little orange strips to overhanging branches at eye level. Every time he tied on a new flag, he glanced back to see how it lined up with the preceding one. As he did this, Branham tried to envision which trees would need to be cut down. He compared this process to playing chess. "You gotta always be thinking seven moves ahead. You're looking at these trees, and you're thinking, *These will be gone. These will be gone. I can weave around this one . . .*"

It struck me that this kind of trail-building was unlike anything else in the animal kingdom: Instead of sketching out a rough line, which would be improved by subsequent walkers, modern trail-builders attempt to find the sleekest route in advance—so that subsequent walkers will never have a reason to diverge from it. In this sense, a hiking trail shares more in common with a modern highway than it does with an ancient Cherokee footpath.

The construction of a trail can appear strangely unnatural, too. Most of the time, trail work involves using a primitive tool called a Pulaski (an axe-adze hybrid) to cut out a narrow, flat trail bed from a hillside, but as a professional trail-builder who spent most of his time working alone on private land, Branham opted for a 2,500-pound machine called a skidsteer, which he used to move soil around until he had achieved the correct grade. Once the trail was finished, he used a leaf blower to cover the trail with plant litter, then, to compact the surface, he would sometimes ride back and forth a few times on his '87 Yamaha BW350 dirt bike.

On a wilderness trail like the AT, the goal of trail-building is, somewhat paradoxically, to artificially create something natural. I witnessed this process up close one summer, while volunteering on a trail-building crew called Konnarock. Here I helped to construct a stone wall, known as "cribbing," to buttress a section of trail across a particularly steep hillside. The wall took three days of backbreaking labor to complete. ("Building a crib wall is like doing a jigsaw puzzle," one of the veteran trail-builders in my crew joked. "Except all the pieces weigh five hundred pounds. And they're all missing.") Once it was finished, we covered the top with dirt and leaves, so that future hikers would scarcely know it was there. This, explained our crew leader, Kathryn Herndon, was the ultimate aim of trail-building: meticulous construction, artfully concealed. One famous trail-builder in Maine, named Lester Kenway, was known to carefully fill in every hole he drilled in a rock face, so that hikers climbing one of his rock staircases could be fooled into thinking some benevolent god had simply dropped the rocks in that arrangement. "The ultimate compliment paid to a trail crew," wrote Woody Hesselbarth, "is to say, 'It doesn't look like you had to do much work to get through here.'"

Benton MacKaye once said, "The Appalachian Trail as originally conceived is not merely a footpath *through* the wilderness but a footpath *of* the wilderness." The same can be said of all wilderness trails: They are both a conduit and a symbolic representation of the wild. Trail-building handbooks invariably stress the importance of maintaining the trail's "primitive character." This is more than a matter of mere aesthetics. There is a crucial difference between a trail that "lies lightly on the land," as trail-builders like to say, and a wide footpath lined with handrails and park benches: the former allows us to experience the complexity and roughness of the world beyond us, while the latter gives us the impression that the world was put here *for* us.

Herein lies the delicate task of the trail-builder: to capture a sense of the wild, to bring order to an experience that is by definition disordered. It is akin to catching a butterfly with one's bare hands. Cup too gently and the butterfly will flutter away, but clap too hard and the butterfly will cease to be.

+

At the top of Smarts Mountain stood a lone fire tower, rising high above the trees. Doyi and I dropped our packs and climbed up the spiraling steel spine of a staircase. At the top, I lifted a heavy wooden trapdoor and we crawled inside. The interior was empty, dusty, enclosed by broken windows. Down below, in every direction, green waves rolled toward the horizon.

Doyi took off his shoes, releasing a swampy, dead smell. "Man, these things are *rotten*," he said. We both hung our socks out the window to dry, while we sat on the wooden floor and ate dried fruit. Doyi sat with his legs outstretched and crossed at the ankles. His feet were a horror. White, wrinkled, and blistered, they would have looked at home on the body of a dead grunt. The toenails on both big toes were plum colored, and his pinkie toenails had already fallen off. He began pointing out others that would soon go: "I'm gonna lose this toenail, this toenail, this toenail, and probably this one," he said. "My feet have *never* hurt this bad before," he said. "Ever."

Five years earlier, on my thru-hike, I had sat in that exact spot, atop the same fire tower, for an entire afternoon, unable to summon the strength to leave. To pass the time, I lay on the floor and listened to a little yellow pocket radio I had purchased to ward off loneliness, but which rarely got clear reception. My body was failing under me. After almost four months on the Appalachian Trail, with only a month left to go, I was a sorry sight. There was scant insulation left on my frame—fat, muscle, or otherwise—with the exception of my legs, where equine muscles flickered and pulsed. I was always wet and cold, and I seemed to have caught some kind of flu back in Vermont. At night: shivers, followed by fever sweats that stunk of ammonia, and then worse shivers. In the morning: more miles to walk. Always, more miles.

Then, without warning, I bumped into my friend Snuggles, whom

I hadn't seen in months. I found her fetused in her sleeping bag on the floor of a damp lean-to; she had been there for three days, lost in a sunless funk. Shortly after we joined up, we ran across another friend of ours named Hi-C. And at last, just as the three of us were entering the White Mountains, the months-long spell of rain broke. The following weeks were sunny, idyllic. Reenergized by good weather and good company, all three of us reached the top of Katahdin one warm morning that August.

I told Doyi this story as we sat atop the fire tower, but it was of no comfort to him. No matter what I said, he still had to put his wet boots back on.

We climbed down the fire tower. I stopped to refill my water bottles from a thin spring that slunk along over the rocky ground. Doyi hiked off ahead, saying I would surely catch up with him. For a long time, though, I didn't. The rain had stopped and the sun again warmed the plants. I savored the spicy air. As I walked, I saw the trail with new eyes; I noticed how water flowed off it, where it pooled, where hikers had tiptoed outside of the trail bed, widening it. At one point during my time working with the Konnarock crew, Herndon told me that being a trail-builder had permanently altered the way she saw trails. "It always takes a few miles, when I go backpacking, to stop analyzing problems and mentally building staircases," she said. "It's hard to stop looking for that stuff, once you've trained your brain to analyze it."

When I caught up to Doyi, he was almost hobbling, his large, green pack swaying from side to side with each step. We descended the mountain at a jerky pace, rather than in the rolling gait—faster and more fluid than a walk but not quite a jog—that thru-hikers tend to adopt.

Doyi talked more about his feet, and home, and missing his grandson. We arrived at the Hexacube Shelter around six. Inside were Doyi's friends, who chatted boisterously as they cooked dinner. Socks had fashioned a balaclava into a fake beard and was doing an impression of a male thru-hiker. The others were toppled over with laughter.

Doyi remained quiet. He cooked two dinners and ate them, back to back, with the air of a man beyond the condolences of food. When the conversation died down, he waited a beat, then said:

"Guys, there's something I have to tell you. I'm thinking about getting off the trail."

They all spoke at once, in disbelief.

"I'm just feeling so weak," Doyi explained. "Climbing up here, I almost fell backwards at one point."

There was a pained pause. Gingko was the first to speak. He too was often dizzy, he said, and his bones hurt. Socks said she had considered quitting once after taking two days off in Virginia, and again in Massachusetts when the mosquitoes were torturing her. Tree Frog began asking Doyi a series of gently probing questions to find out what the source of the problem might be. He asked him what he regularly ate (buffalo jerky, dried fruit, mac and cheese, oatmeal) and in what quantities (not nearly enough). Tree Frog suggested that Doyi buy food that was more calorically dense—a good rule of thumb was that any food worth carrying should have one hundred calories for every ounce it weighs, he said. Each of the hikers began suggesting foods that fit this criteria: peanut butter, olive oil, summer sausage.

They began fishing items out of their backpacks and handing them to Doyi. Tree Frog gave him some trail mix and a packet of electrolyte powder. Gingko chipped in a candy bar. With a solemn air, Catch-Me carried over a black package of ginseng—the real stuff, he said, very expensive—and a small bag of pink rock salt. Someone offered Doyi a jar of peanut butter, but he politely waved it off. He said he already had one, and lifted up a sixteen-ounce jar.

"How long has that lasted you?" Tree Frog asked.

"About two weeks," Doyi said.

"I eat one of those in *two days*," Tree Frog said. He suggested

Doyi keep the jar on top of his pack, and anytime he stopped, for any reason, to stop and eat a spoonful. After a few minutes more of this, I pointed out that by the time these guys were done with him, they'd have to change his trail name from Doyi to Doughy. Doyi laughed. His gloom had lifted slightly.

Next, Doyi's friends started going through his pack and suggesting things he could leave behind: a trowel (meant for digging cat holes, but generally extraneous, since any old stick works almost as well), a large bottle of tick repellant (too big), a Nalgene water bottle (too heavy), a water filter (could be replaced with two tiny bottles of chlorine dioxide solution), a blue tarp (could be swapped for a piece of Tyvek home wrap, or ditched altogether), a knee brace he never used, spare clothes, spare shoes. Tree Frog even offered to send his tent home so he and Doyi could split the weight of Doyi's two-man tent. This was a generous offer, because logistically, it would lash the two of them together for the rest of the trip, hell or high water.

Watching Doyi's fellow thru-hikers come to his aid, it occurred to me how remarkably humane a space the AT has become, compared to most wilderness footpaths. Some hikers deride the AT as a mere "social trail," as opposed to the wilder, lonelier trails out west. But I imagine Benton MacKaye would have borne that label proudly. His original intention was not just to give people an escape from urban environments; he wanted to set aside a space where people could unite around the common effort of living outdoors, a place where "cooperation replaces antagonism, trust replaces suspicion, [and] emulation replaces competition." The trail that eventually grew out of that vision wasn't utopia, exactly, but it was a start.

"If you want to finish, we'll do whatever we have to in order to get you there," Tree Frog said.

Doyi thought a moment. He made a small, pained smile.

"I do," he said, firmly.

"We'll get you there," Tree Frog said.

Doyi thanked him.

Tree Frog shrugged. "*Nigada Osda,*" he said.

At the time, I mistook that phrase for the group's rallying cry: *Osda Nigada,* "It's all good." In fact, Doyi later told me, some weeks after returning home from the windy summit of Katahdin, what Tree Frog had said was another Cherokee phrase: *Nigada Osda.* "Everybody is good."

CHAPTER 6

THE IDEA to radically lengthen the Appalachian Trail occurred to Dick Anderson one afternoon in the fall of 1993. He was driving north through Maine on Interstate 95, a major highway that runs the length of the East Coast and dead-ends at the border of Canada. As his eye followed that north-south line, his mind made a parallel hop. Anderson knew that the Appalachian mountain range continued north past Katahdin and ran up along Canada's east coast, before slumping into the ice-clotted North Atlantic. Why, then, he wondered, couldn't someone extend the trail into Canada?

He had no idea where the idea came from—he had never hiked a single mile of the Appalachian Trail. It was, he later recounted, as if his mind's antenna accidentally intercepted a message intended for someone else. *Holy shit!* he thought. *How come no one ever thought of this? This is a wicked idea!* He pulled over to get gas, and, impatient to share his plan, he began explaining it to a man at an adjacent pump. "Of course, he's over there looking at me like I'm freaking *nuts*," Anderson recalled.

Anderson, a former commissioner of the state's Department of Conservation, went home that night and unfurled a regional geophysical map—one virtually devoid of towns, roads, or borders—and began laying a line of little blue sticky dots along the ridge of the Appalachian range, linking the highest peaks in Maine, New Brunswick, and southern Quebec. Furtively at first, he began showing the map around and gauging his friends' and colleagues' reactions.

When he finally made his proposal for the International Appalachian Trail public, on Earth Day 1994, representatives from New Brunswick and Quebec quickly agreed to the extension. Over the next few years, he began receiving calls from representatives of the Atlantic islands of Prince Edward Island and Newfoundland—places that also possessed Appalachian geology—urging him to lengthen the trail even farther. He eagerly said yes to each. Of course, hikers would have to ride on a ferry or an airplane to reach these islands, but, Anderson thought, so what?

Shortly after the International Appalachian Trail committee approved the Newfoundland extension in 2004, one of Dick Anderson's friends, Walter Anderson (no relation), a former director of the Geological Society of Maine, began circulating a map showing that, in fact, the geological Appalachians continued on the far side of the Atlantic Ocean, in a more or less mirror image of the North American range. Some four hundred million years ago, he explained, the continental plates began to collide, forming the Pangaea supercontinent. That slow-motion crash lifted up the ancient Appalachians to heights rivaling those of the modern Himalayas. But when Pangaea broke apart two hundred million years later, the continents that would become North America, Europe, and Africa split along that raised seam, like a piece of paper folded and torn. For this reason, Appalachian rocks can be found scattered throughout the soil of Western Europe and North Africa.

Dick Anderson was smitten with this notion. If the Appalachians continued all the way to Morocco, why stop in Canada? What was holding the trail back? A few (admittedly, sizable) bodies of water? A few (okay, most) people's antiquated notions of what a trail should look like?

At the time, the full reality of what he was proposing—the daunting task of blazing and maintaining the world's longest hiking trail—was still far off. But Anderson, like Benton MacKaye, intuitively understood that the task of creating a super-long trail principally consisted not of trail-building but trail-linking. The artistry lay in the elegance of the connections, the tightness of the joints, the sinuosity of the curves, and, more than anything, the strength of the idea that would hold them all together—what Anderson referred to as the trail's "philosophy." In those years between 1993 and 2004— the brightening dawn of the Internet Age—it was only natural that the big idea undergirding Anderson's trail, when it came to him, was *connection*: of people, of ecosystems, of countries, of continents, and of geologic epochs.

+

Certain trails are so elegant that they seem to lie sleeping just beneath the surface of the earth. Rather than being created by us, it is as if these trails unveil themselves *through* us. When humans, bison, deer, and other woodland animals go in search of the shallowest pass in a mountain chain, they tend to decide on the same route. Who, then, invented the trail? The humans? The bison? The deer? The answer, it seems, is that no one can claim full credit, because an essential trail— a path of least resistance—is predetermined by the shape of the topography and the needs of its walkers. Just as biologists sometimes say that "function precedes structure," in some sense, a trail precedes the trail-maker, waiting there for someone to come along and brush it off.

Brilliant technological innovations, according to the tech philosopher Kevin Kelly, are created in the same seemingly inevitable way.

For instance, once humans had invented the road network, the horse-drawn carriage, the internal combustion engine, and a fuel like gasoline, it was only a matter of time before someone synthesized them into an automobile. It is no coincidence that Karl Benz and Gottlieb Daimler independently created the modern automobile within a year of each other (and that several other inventors created their own variations within a few short years). Once there is a use for a technology and the right components exist, inventors simply need to make the right connections. This rule applied in turn to each of the technologies that made up the automobile: the engine, the metallurgy, the wheels. Each was, in retrospect, an inevitable shortcut across the intellectual landscape, which then allowed for future shortcuts.

Viewed in hindsight, it can appear that great trails and great inventions are both preordained. But Kelly is careful to point out that while various forces can create the right conditions for a given technological breakthrough, the final form that technology will take is not predestined; any new invention is still profoundly shaped by its inventor. The incandescent lightbulb, for example, was invented by twenty-three separate men, each of whom imbued the same basic mechanism with his own unique shape and design. Kelly likened this interplay between inevitability and serendipity to the formation of snowflakes, which unfurl into unique existence when a seed (usually a mote of dust) encounters the right environmental conditions (a supersaturated, supercooled cloud). "The path of freezing water is predetermined," Kelly wrote, "but there is great leeway, freedom, and beauty in the individual expression of its predestined state."

When Benton MacKaye first proposed the Appalachian Trail, the conditions were in place for the birth of a new kind of hiking path: hikers were walking farther; trails were growing longer; and planners were thinking on a grander scale. In fact, by the time the AT was first proposed, there had already been numerous proposals for long trails to stretch the extent of the Appalachian range. "The one big supertrail," wrote Guy and Laura Waterman, "was inevitable." However, Benton MacKaye's proposal, with its inspiring rhetoric about wilderness preservation and the plight of the working class, was the formulation that ignited the public's imagination. Once MacKaye proposed it, the trail burst into being.

In 1993 Dick Anderson seemed to have also stumbled on that golden thing—an unrealized inevitability. At that precise moment, a hunger was growing in the world for longer and longer trails. The shift toward monumentalism had begun with the Appalachian Trail; then, in the 1980s and '90s, trails like the Pacific Crest Trail and the Continental Divide Trail sprang up and outgrew it. Supertrails— hiking paths measuring more than a thousand miles—started being built in Russia, New Zealand, Nepal, Japan, Australia, Italy, Chile, and Canada. Much of Western Europe was also webbed with supertrails, like the famed Grande Randonnée network of walking paths. In part, this growth was fueled by the availability of ever lighter hiking gear, which allowed people to walk greater distances. As long-distance hiking grew in popularity, long trails became more crowded. By the turn of the twenty-first century, some thru-hikers had begun to complain that supertrails like the AT had lost the lonely, wild quality that originally made them alluring. The conditions were ideal for a radically longer long-distance trail to be born.

Virtually as soon as Anderson proposed extending the IAT overseas, the proposal took hold, with Scotland and Spain both expressing interest in 2009, and most of the other countries following close behind. Much of its route already contained trails that were simply waiting to be stitched together. Where the IAT crossed over from Northern Ireland to the Republic of Ireland, Anderson encountered contiguous trails that were managed by different (and mutually unfriendly) trail groups. Connecting them required no work at all— only a paradigm shift. When I first met Anderson, in Portland, Maine, in the spring of 2011, he had just learned that the maintainers of the

North Sea Trail, which passes through seven countries in Northern
Europe, had voted to join the IAT. "That's six thousand miles right
there!" he said. "Schwoop! Cross that off the list."

When this idea first occurred to him back in 1993, Anderson
had no way of knowing that it would grow so smoothly or so far. In
fact, early on it had looked like the plan might face fierce opposition.
Not long after his great brainstorm, he had showed the map covered
in blue dots to his friend Don Hudson. "Dick, this is a great idea,"
Hudson had replied. "And they're going to *hate* it." Hudson was referring to members of the Appalachian Trail Conservancy and the
managers of Baxter State Park, both of which resisted the idea of
extending the AT into Canada, because that would entail blurring
out the near-sacred terminus of Katahdin. (Even two decades later,
the word *extension* remained so taboo that Don Hudson referred to
it as "the e-word.") The AT and IAT factions eventually reached a
fortuitous compromise: instead of an "extension," Anderson opted
to instead call the IAT a "connector trail." The IAT would *connect* to
the AT, and in turn connect the AT to the world.

+

The core function of any trail is to connect. The root of that word,
from the Latin *connectere*, means to "bind together" or to "unite." In
this sense, a trail strings a line between a walker and her destination,
uniting the two in an uninterrupted corridor so that the walker can
reach her end swiftly and smoothly. Since the rise of electrical engineering in the nineteenth century, a second sense of the word has
gained widespread use. When two things remain distant, to connect
them means to create a conduit through which matter or information
can flow. Here again, trails act as connectors: when a trail is blazed
between two towns, a line of communication is established; people
can travel back and forth, goods can be exchanged, and information
can spread.

Humans and other animals have long used trails to link the essential loci of our environments. The brilliance of trails is that, over time, they naturally streamline to reach their goals faster or with less effort. Like elephant trails, humanity's footpaths eventually grew taut along the landscape's paths of least resistance. However, efficient as these connections may have been, even the best trails had a speed limit: walkers could only reach the trail's end as fast as their legs could carry them. So our next impulse would have been to train ourselves to run faster. Larger societies—dating back at least to the ancient Sumerian city of Uruk—designated a specialized class of running messengers, who could transmit our messages even faster across long distances.[*] In many empires, new kinds of paths were built to accelerate the flow of messengers. These advanced footpaths reached their apogee in the Inca Empire, where trails were paved with flat stones and equipped with staircases, shade trees, bridges, rest huts, and watering holes. Along these paths, imperial messengers ran relay-style, six miles at a time, while passing along knotted strings called *quipus*, which bore simple (often numeric) messages. In this manner, information could move about one hundred fifty miles a day.

Everywhere that people wanted to go faster, our trails grew straighter, flatter, and harder than ever before. What set humans apart from our animal brethren is that we learned to optimize *beyond* the shape of the trail and the limits of our anatomy; technology, in a sense, provided an entirely new dimension in which trails could streamline. To travel and transport goods faster, people in Eurasia discovered that they could ride atop animals and hitch them to carts.

---

[*] An ancient Sumerian poem tells that the very first text was written by the legendary King Enmerkar, who wanted to send a message across the mountains to a rival king, but found that "the messenger's mouth was too heavy and he could not repeat it," so he carved it onto a clay tablet. (The rival king, it's said, could not interpret the message, but was so awed by the new technology that he was forced to surrender.)

(Domesticated animals, in this way, became a kind of living technology.) Roads adapted in new ways to the technology of wheeled transport; in ancient Babylon, they built "rutways," stone roads bearing parallel grooves to guide the wheels of bulky carts, which were an early precursor to wooden and then metal railways. Generation after generation, Eurasians continued to improve their vehicles and roads, until they invented the automobile, or "horseless carriage," and the locomotive, or "steam horse." Soon, using these machines, humans were racing across the land faster than any animal on any trail. But even that was not fast enough. So next, like Daedalus, we fabricated wings and taught ourselves to fly.

As we discovered new ways to make our bodies travel faster, we also learned to send information at even more astonishing speeds. Early communication technologies like smoke signals and drums encoded simple messages in visible and audible forms, allowing people to transmit information across long distances nearly instantaneously. The invention of electricity allowed for yet more complex messages to be sent even farther. That shift began with the invention of the telegraph, which was followed by the telephone, the radio, the television, the fax machine, and eventually the computer network. Today, information constantly spirits past us, a ghostly chatter between billions of people and machines; our connections have sleekened so drastically and spread so far that they've effectively vanished from sight.

A trace, when followed, becomes a trail. Likewise a trail, when transformed by technology, becomes a road, a highway, a flight path; a copper cable, a radio wave, a digital network. With each innovation, we're able to get where we want to go faster and more directly—yet each new gain comes with a feeling of loss.

From trains to automobiles to airplanes, each time the speed of connection quickens, travelers have expressed a sense of growing alienation from the land blurring past our windows. In the same vein, many people currently worry that digital technology is making us less

connected to the people and things in our immediate environment. It
is easy to dismiss these responses as overreactions, the curmudgeonly
groans of the progress-averse. Yet in all these cases, a faster connection
palpably diminishes our ability to experience the richness of the physical world: A person texting with her friends or riding on a bullet train
is connecting very quickly to her ends, but in doing so, she skips over
the immensely complex terrain that lies between those two points.
As the anthropologist Tim Ingold has pointed out, instead of being
immersed in an endless continuum of landscapes, we increasingly
experience the world as a network of "nodes and connectors": homes
and highways, airports and flight routes, websites and links.

The importance of place and context—those two words whose
meanings twine in the word *environment*—necessarily wanes as we transition to a world of nodes and connectors. The fact that trails enable just
this kind of reduction in complexity has always been one of their chief
appeals. But the faster we travel, the more intensely we feel our lack of
relationship with the land we traverse. And so, beginning roughly with
the advent of locomotives, new trails were built for the very purpose
of reconnecting us back to (and, later, preserving) the environment.
These trails webbed together, and lengthened, until one could walk
from one end of the country to the other, remaining almost always
within a wild landscape, where (in the memorable language of the
1964 Wilderness Act) "man himself is a visitor who does not remain."

It can be hard to see exactly where the IAT—the great connector—
fits into the grand history of trails. Is it the continuation of a trend?
A return to a prior mode? Something wholly novel? To answer this
question, it helps to first ask: What desire is this trail fulfilling? As I
traced the IAT from Maine to Newfoundland to Iceland to Morocco,
I began to realize that the IAT seeks to resolve our confused feelings
about scale and interconnection. The trail itself is a surreal project:
standing on a mountaintop in Scotland, you somehow understand
that this is the same mountain chain you once climbed, years earlier

and an ocean away, in Georgia. In an era when we are able to travel through the air with godlike nonchalance and send information to other continents at the speed of light, a truly global footpath confirms our belief in how connected, how small, the world has become, and yet it also reminds us how unfathomably—how *unwalkably*—huge the planet remains.

+

In the fall of 2012, I traveled back to the summit of Katahdin and began walking north toward the border. I was equipped with a set of instructions and a map I had printed out from the IAT website, which gave me turn-by-turn instructions to lead me through the concatenation of forest paths and roads that made up the trail. I knew it would take me roughly a week to reach the border, but I had no idea what lay ahead of me. When you set out to hike the AT, you carry with you some sense of what the storied Long Green Tunnel will entail. I bore no such preconceptions about the IAT. It was terra incognita.

I tiptoed along the Knife's Edge, over Pamola Peak, and then down the eastern flank. From the base of the mountain, I trudged down a wet gravel path and skated along algal-slick boardwalks. The cold air dripped. I wore three layers (merino, synth puff, rain shell) and a winter beanie, and still I shivered. It was October, and the leaves were in full death-bloom. A small frog, spotted like a leopard, flopped out of my path on drugged legs.

The route turned onto an old overgrown carriage path and continued until it reached a gate that led to a wide logging road. On my left was a brown wooden sign indicating the "southern terminus" of the International Appalachian Trail, painted in the same white-lettered, hand-carved style of those on the AT. Below it was the IAT blaze: a white metal rectangle, about the size of a dollar bill, surrounded by a blue border. Printed onto the white background were the cruciform letters:

S I A
A
T

I was officially on the IAT; this was the first of hundreds of thousands of blazes that would one day mark the trail from here to Morocco.* At around twelve thousand miles, it was a project on a truly
planetary scale: if a hole were dug from Hawaii to Botswana through
the Earth's core, one would have to walk all the way to the end of
that Hadean tunnel and halfway back in order to travel an equivalent
distance. The length is so staggering, and the climates so punishing,
that most people doubt any person will ever walk its full extent in one
continuous trip. Anderson told me he was doubtful as well, but he invited people to try. After all, he pointed out, people once thought the
Appalachian Trail was too long to thru-hike. In 1922 Walter Prichard
Eaton, an early architect of the trail, predicted that "The Appalachian
Trail would exist in its entirety chiefly for a symbol—that is, nobody,
or practically nobody, would ever tramp more than a fraction of its
length." When Earl Shaffer completed the first thru-hike of the Appalachian Trail in 1948, the ATC initially greeted his announcement
with skepticism. "But the fact is that he made the Appalachian Trail
work," Anderson said. "You only have to have a few people walk from
one end to the other to express the purpose of your trail."

Back when I was hiking the Appalachian Trail, I had met a rusty-
bearded fanatic named Obi who told me that once he had reached
Katahdin, if all went according to plan, he would continue hiking an
additional eighteen hundred miles on the IAT up through eastern
Canada to the northern tip of Newfoundland. He had gotten the idea
from a famous thru-hiker named Nimblewill Nomad, who, in 2001,

---

* SIA stands for "Sentier International des Appalaches," a concession to the
Quebecois, whose law mandates all signs must be displayed in French.

became the first person to hike from the southern tip of Florida to the northern tip of Newfoundland, covering some five thousand miles. I looked upon these fanatics with mingled reverence and suspicion, as I would a gourmand who had decided to top off an enormous steak dinner by slurping down three dozen oysters. (*Wasn't the Appalachian Trail long enough?* I thought. *Why keep going?*) But out here, all alone, I caught a glimmer of the feeling these super-thru-hikers were chasing. It was the same feeling the early AT thru-hikers must have experienced: lonesome, uncertain, faintly electric. It felt like adventure.

Northern Maine, late fall: even the sunlight has a dark, ice-whetted edge. The trail followed a wide logging road for five miles, which undulated through stands of Jupiter-toned second-growth forest. Over the next few days, the logging road turned into a vanishingly faint dirt trail, which then turned into a riverside tow-path, then a dirt road, and then—with the exception of a maddeningly straight stretch of bike path, a set of ski runs, and a surreal section in which, for a distance of eight miles, it *became* the US-Canadian border*—the trail ran along paved roads until it crossed into Canada.

Once it hopped the border, the IAT became stranger still, jumping from one Maritime island to another, brazenly flaunting the notion of contiguity. Farther north, in Newfoundland, the trail sometimes split into multiple routes, or it disappeared altogether, forcing hikers to navigate with a map and compass, such as in the Tuckamore-choked section I would hike on the west coast of Newfoundland. A trail is traditionally defined as a single, walkable line. But this new, slippery, sprawling, leviathanic thing—which swallowed roads, leaped seas, and vanished from sight—was quietly redefining the term.

---

* This trail, I was informed by a border patrol officer, was known as the "boundary vista"—a twenty-foot-wide swath of cleared land that constitutes a geopolitical gray area between the two nations. How fitting, I thought, that the International Appalachian Trail had co-opted a purely international space.

I dislike walking on roads, so whenever I encountered concrete, I stopped and stuck out my thumb. Sometimes I had to wait for an hour or two, but a car would inevitably pull over and pick me up. Then we zoomed away down the long, straight farm roads, covering days of walking in an hour. I must confess: it felt like magic. But I did miss the views walking affords. Staring out through the windshield and the passenger-side window, I saw much of the Maine countryside in a series of freeze-frames and blurred pans, that weird wave-particle duality familiar to all car passengers. We passed maple syrup stores, potato farms, Amish men on bicycles, and old barns, hollowed and phantasmagorically warped, but still, somehow upright.

In the areas where the trail diverged from the road, I resumed walking. In the process of hopping in and out of cars, I was forced to pay new attention to the oddly cyborgic nature of travel in the industrialized world. On a vacation to a foreign country, a person might unthinkingly use a half-dozen different modes of transportation—we walk, we drive, we fly, we ride on trains or streetcars, we sail on ferries, and then we walk some more. On the IAT, I began to notice how many other machines quietly aided in my survival: not just the cars that carried me, but the heavy machinery that paved the roads and bike paths I walked on, the computers that printed out my map, and the factories that built my gear. I ate food cooked and dehydrated and packaged and rehydrated and recooked with the use of machines. At night I slept in strangers' homes (machine-built, machine-warmed, and filled with smaller machines), or in a wooden lean-to (whose materials had been trucked or choppered in), or, one night, directly beneath the spinning white blades of a wind farm.

The oddest part, on further reflection, is that all this technology seemed utterly normal, even *natural*, to me. This deep and often unconscious reliance on technology has inspired the design theorist and engineer Adrian Bejan to dub us the "human and machine species."

Humans adapt to their environments. One of the ways we adapt is by creating technology. Once an invention is widely adopted, it effectively becomes part of the landscape, another feature to which our lives adapt. We then create more technology to adapt to the existing technologies. A smartphone, for example, is adapted not just to human anatomy and the physical constraints of the earth, but also to a network of cellular towers, a constellation of satellites, a standardized system of electrical jacks, a wide variety of computers, and a telephone system stretching back to the middle of the nineteenth century, which was strung together with wires made from copper, a substance we have been manipulating for the past seven thousand years.

Our innovations pile up, one atop the other, each forming the foundation for the next, until an entirely new landscape, a *techscape*, emerges—like a city built on the ruins of past empires. Any person who tries to resist the adoption of a vital new technology begins to feel this transformation acutely; Luddites become, quite literally, maladapted to the modern world. For instance, I swore for years I would not buy a smartphone, because they seemed unnecessary and expensive. But then my friends started texting me videos or web links, which my cheap flip phone couldn't open; and the fact that I would need to look up addresses and directions before leaving the house became a handicap, as GPS made it easy for others to make plans on the spot. Finally, half to keep from falling out of touch, half to keep from getting lost, I broke down and bought a smartphone too.

In this techscape, new values also emerge—often made up of old words with new connotations: *automatic, digital, mobile, wireless, frictionless, smart*—and new technology adapts to those values. The current meaning of the word *wilderness*, one could argue, emerged directly from the techscape of industrialism, just as the current meaning of the word *network* emerged from the world of telecommunications. With the advent of industrial technology we began to see wilderness less as a landscape *devoid of* agriculture and more as a landscape *free from* technology—and thus the wild went from being a wasteland to a refuge.

Much of our modern conception of wilderness was formulated in direct opposition to the technology of mechanical travel: William Wordsworth, Britain's foremost nature poet, preceded the modern environmental movement by a century when he vociferously opposed the expansion of a proposed train line into the Lake District of Northern England. In the United States, both Aldo Leopold and Bob Marshall defined wilderness as an area (in Marshall's words) that "possesses no possibility of conveyance by any mechanical means." Benton MacKaye agreed; much of his later life was dedicated to fighting the incursion of "skyline highways" into the Appalachians. Together, these three men, along with a handful of others, founded the Wilderness Society.

In large part, the continued interest in hiking seems to stem from a desire to cut through the techscape to get to some *natural* substratum: to borrow MacKaye's phrase, to see the "primeval influence" beneath the "machine influence." But ironically, the act of hiking is also dependent on technology. Many of the earliest hikers relied on trains and automobiles to reach the mountains. Today, some forms of technology (like cell phones or ATVs) are considered obnoxious, while others (like water purifiers, camp stoves, and GPS locators) are excused. In either case, technology inexorably trickles into the wild, allowing hikers to reach new lands, travel in new ways, think in new terms, and optimize to new values.

Wilderness looks different in the neon light of technology. In the traditional framework of wilderness preservation, a techscape is merely a despoiled wilderness landscape. But when viewed through the lens of technology, the wilderness can be seen as nothing more than an ultra-minimalist techscape designed to provide an escape from other, more baroque techscapes. Readers raised on the wild gospels of John Muir and Edward Abbey will likely cringe at this definition—as, indeed, I once did. Such is our aversion to mixing technology and wilderness, even in theory. But while walking the IAT, I came to appreciate the matter with a bit more nuance. While most trails try to hide their fraught relationship with technology—by banning motorized transport, avoiding roads, disguising their own construction, and in all other ways, aping primordial nature as best as possible—the IAT bears it unabashedly, like a smiling mouth full of gold teeth.

+

Following the IAT north, I flew to Newfoundland and hitched more rides. For most of its length, the Newfoundland section follows a highway up the island's west coast called Route 430, which an enterprising local tourism association had dubbed the "Viking Trail." Beginning in Deer Lake, I hitched my way from small town to small town, stopping off here and there for scenic hikes. (This combination of long drives and short hikes is precisely how the trail's architects had envisioned most people using it. Thru-hikers were manifestly not their primary concern.) Eventually, I made my way to the trail's northernmost point, a place called Crow Head. There, I walked along a gravel footpath for less than three miles before I reached a bluff overlooking a wide expanse of ocean dolloped with icebergs. I found no sign marking the trail's end. Or rather, there *was* a sign, but, as I would later learn, the sea winds had scoured it blank. I hung around on that bluff for a long time, trying to imagine how it would have felt to stand there, triumphant, after walking all the way from Georgia.

I felt no sense of triumph, not even secondhand. Instead, what I felt was something like guilt or loss. By having hitchhiked there, I had cheated myself of the slow engagement with the local landscape that thru-hiking provides. Hitching was too easy, too quick.

There was one upside to hitching, however: it was a wonderful way to get to know the local people. Snug in the confines of the car, staring ahead at the road, conversation naturally flows between drivers and passengers. In an amazingly short span of time, awkwardness, suspicion, and fears of impending murder give way to a form of intimacy resembling that of a second date. In Newfoundland, I caught rides with fishermen, miners, carpenters, and, once, a trucker hauling a load of recyclables whose eighteen-wheeler had a pair of red-stained moose antlers bolted to the grille. (He told me that on average he inadvertently ran over about twenty moose a year.) The drivers were exceptionally generous. They were constantly offering me beer and drugs—a welcome inversion of the traditional ass-gas-or-grass economics of hitchhiking. In return, all they wanted was someone to talk to.

At some point in our conversations, I managed to ask every one of these drivers if they had heard about the IAT, the trail they were driving on at that very moment, which would one day stretch all the way to Morocco. None had ever heard of it. A few of the drivers said they had noticed gaunt men and women carrying backpacks and "ski poles" along the side of the highway, though none knew where, or why, they were walking. On these long stretches of the IAT, drivers and thru-hikers shared the same route, but they were in two distinct landscapes—the land of the slow, and the land of the fast. It is strange, then, but also strangely appropriate, that Dick Anderson should have first dreamed up the IAT while driving along the highway. As I would later learn, hiking trails and highways, like the snakes on a caduceus, have always been both opposed and curiously entwined.

+

Most people would be surprised to learn that the American interstate highway system, as it currently exists, was first envisioned by the AT's founder, Benton MacKaye. In 1931, MacKaye (along with his friend, the forester Lewis Mumford) proposed the notion of what he called the "townless highway" to remedy the problem of high-speed traffic and congestion passing through downtown streets. The heart of the problem, as MacKaye saw it, was that the road network had evolved from ancient footpaths, which grew into bridle paths and wagon roads. Motorcars were a wholly different technology, with different abilities and limitations, and so deserved a fresh system suited specifically to them. He proposed that cars be given their own dedicated spaces in which they could reach maximum speed, just as trains had been.

Thanks in large part to the Federal-Aid Highway Act of 1956, major highways now regularly skirt cities, and many are bordered not by rows of tacky shops (what MacKaye called "motor slums") but by strips of forest. Cars got faster, towns got quieter, and the circuitry of civilization reorganized itself around a new mode of transport. Quicker than a horse, more flexible than a railcar, the highway-bound automobile was ideally suited to a sprawling, mobile, individualistic nation like postwar America. Now, despite a growing awareness of the automobile's many downsides—namely its inefficiency, pollution, and tendency to kill people—it remains our de facto mode of transportation, in part because everything else in the American landscape has hardened around it.

Though the modern interstate is a recent invention, the history of the highway stretches back thousands of years. When the earliest footpaths were widened into roads, the next logical step was to make those roads faster. Often this involved artificially hardening the road's surface and raising it above the surrounding land so it could shed water (hence the name *high*way). Unlike trails, highways required a massive expenditure of labor to build, so they were only built when rulers were able to marshal the necessary labor force (usually made up of slaves and soldiers). As a result, the earliest highways served as the tentacles of grand empires. Through them principally flowed three things: royal information, royal armies, and royal personages. In imperial China, wide highways (*chi dao*) of finely tamped earth, lined with shady evergreens and paved with flat stones, were built with ruts conforming exactly to the axle length of the emperor's carts and carriages; on each of these three-lane highways, the center lane was reserved for the exclusive use of the imperial family. Along the Roman *via publicae* were installed milestones, which regularly reminded one of the distance from—and thus, the reach of—imperial Rome. When the Assyrians conquered a new region, they built new roads to more quickly transport military dispatches and allow troops to quash local revolts. The Maya did the same. Incoming Inca emperors would sometimes command their conscripted laborers to build new paved roads even if passable stretches of road *already* existed, simply to signify their control of that land.

In colonial America, the evolution of the road system mirrored that of the nation. At first, new European settlements were relatively ungoverned, and paths were rough. Over time, when settlements became sufficiently populous, the government extended its reach, offering legal protection while extracting taxes. The tax system was designed to create more roads: by the 1740s in North Carolina, every taxable male was expected to perform as many as twelve days of roadwork each year, though wealthy people avoided this obligation by paying others to serve in their stead. Unpaid taxes could be recompensed by doing yet more roadwork.

The road network had to balance a need for expediency with a need to connect population centers, so it would often deviate from the path of least resistance in order to accommodate a large town or city. Geographers call this phenomenon "population gravity." However, a more apt term might be "capital gravity," in the sense that the roads, built with tax dollars, bent to service the largest sources of funding. As a general rule, in the colonial era, every publicly maintained road contained at least one large house alongside it, because that household had sufficient influence to sway the government into building a road there. In later years, this rule would hold true, but instead of big houses, the new roads led to big corporate interests, like Alaska's Dalton Highway, which was built by the oil companies in just five months in 1974, with the help of government engineers and funding, to reach the Prudhoe Bay oil fields and service the Trans-Alaska Pipeline.

"Where roads go tells you where the power is at any given time," Tom Magnuson, an expert in the history of the colonial road network, once told me. We were driving around Hillsborough, North Carolina, and he was pointing out the spectral remnants of colonial roads in the nearby woods. The reason we could still locate bygone roads like these, Magnuson explained, was somewhat unintuitive: there are fewer official roads today than there were a hundred years ago. "It's a result of the increased weight of our carriage," he explained. "We're carrying heavier cargo, so the road surface has to be better, more expensive. The more expensive the road, the fewer roads you build." For example, he said, today truckers all take a single route from Raleigh to Atlanta: Interstate 85. It would be slow and costly to go any other way. However, back in 1950, there were about twelve different routes that trucks regularly used. Fifty years earlier, there were probably double that amount.

As they are woven into the fabric of civilization, highways effectively become part of the landscape, which people must then adapt to. Businesses and, in some cases, entire towns (once called "pike towns") spring up to service the highways' passengers. Magnuson had found thirteen hamlets in North Carolina alone that had disappeared after they were bypassed by highways. Other towns, like Timberlake, had migrated to be closer to a major thoroughfare, leaving behind the husks of abandoned neighborhoods. The same phenomenon takes place whenever a fixed structure becomes an essential part of peoples' lives: even if it is initially built to serve us, we end up molding our behavior around it.

From a walker's perspective, the brutal nature of modern highways stems from the ways they have adapted to the technology of the automobile. Because cars have trouble turning at high speeds, highways must seek the straightest line they can, even if that means blasting a tunnel through a mountainside. Because driving fast is much more dangerous than walking, highways inevitably require regulations and penalties—construction codes, speed limits, and, those most dreaded of all creatures, traffic cops. And because a speeding car will kill just about any living thing it runs into, highways marginalize and imperil any human being or animal walking along them.

In this way, highways have created a wholly new, highly technological landscape of movement. Optimized to the "human and machine species," this landscape is maladapted to the naked human body, even though it is the full expression of a deeply human desire—to go farther, faster; to connect in heretofore impossible ways.

The decision to route a hiking path over a highway might seem a deeply counterintuitive one. It is not, however, uncommon. Before automobiles pushed hikers into the hills, walkers and wheeled vehicles shared the roads; in fact, road walking was a popular American pastime in the late nineteenth century. The public's lingering fondness for the concept of road walking would prove crucial to the creation of the AT, which just like the IAT today, originally routed much of the trail over (dirt) roads. This tactic had a clear rationale: in the race for federal funding, the trick to building a long trail is to first create a walkable route and attract attention. Then, once the trail is well known and funding begins pouring in, one can worry about shifting it off roads and into wilder lands, mile by mile. "For example in Maine, the Appalachian Trail once made extensive use of logging roads," Dave Startzell, the former director of the ATC, told me. "Over time we were able to relocate a lot of those sections. But you're talking about a program that spanned more than thirty years, cost more than two hundred million dollars, and involved acquiring over three thousand parcels of land."

There is a catch-22 inherent to this process, though. Trails need money to relocate away from roads. In order to raise money they need to attract hikers (and media coverage) to demonstrate that the trail is desired by the public. However, most hikers dislike walking on roads. A similar logical bind commonly arises with the creation and adoption of any new technology: With highways, for instance, if everyone in a given region were to pitch in (with their money or their labor), a new highway could be built very quickly; but it is very hard to prove that the highway is necessary or useful until it has been completed and people are already using it. In this regard, the hikers who had committed themselves to hiking the IAT now, when it is ugly and hard, were committing something akin to an act of faith. They were walking the trail into existence.

After returning home from Canada, I tracked down two such thru-hikers, Warren Renninger and Sterling Coleman. In 2012 they both successfully hiked from the southern tip of Florida up to the northern tip of Newfoundland along the IAT (a trip of some five thousand miles). Coleman told me he enjoyed how the road walks connected him to local people—he was often offered food, and on one occasion, a pack of boys chased him down to ask for his autograph— but overall he found it an alienating experience. The long, straight stretches of road played tricks on both their minds. Renninger said that if he had it to do over again, he would have brought along a bike to make these sections go by quicker. Coleman said that, since many of the roads were never blazed, over time the trail seemed to dissolve before his eyes. "It started to feel more and more like I wasn't actually on a trail," he said. "It just felt kind of arbitrary that I had a piece of paper that said I had to go here, here, and here . . ."

+

Somewhere around Iceland, things began to fall apart. In the spring of 2012 I received word from Dick Anderson that the IAT would be holding its first overseas general assembly in Reykjavik, a historic event, designed to cement the ties between the scattered branches of the organization. He warmly invited me to attend, and so, using the remaining funds of a fellowship I'd been awarded, I booked a ticket.

I landed in Reykjavik at midnight, underneath an apricot sky. It was late June, close to the summer solstice, when the dark of night lasts just three hours. I slept restlessly, and then woke with a start at two P.M. the following day. Already late, I brushed my teeth and washed my face with the sulfurous tap water, then hastened to a reception that was being held for the IAT committee at the US embassy.

The inside of the ambassador's home was awash in timeless sunlight. Gray-haired men and short-haired women held glasses of ice water and small plates ravaged of hors d'oeuvres. I greeted Dick Anderson, who wore a brown corduroy jacket and a tie. Don Hudson introduced me to the chairperson of the IAT, Paul Wylezol, a pale Newfoundlander with a dark Caesar haircut. Wylezol was a stern, saturnine presence, with an odd habit of referring to the IAT as a "brand": at one point, he said that the routing of the IAT over pre-existing trails, like the ancient Ulster Way in Ireland, wasn't "rebranding," but "another level of branding." (I also overheard him compare the IAT to McDonalds.) In his spare time, Wylezol toiled over a dense tome of deductive logic, which he hoped to one day publish not on paper, but in a sprawling, nonlinear, hyperlinked electronic document. I had long wondered who had given the IAT its curiously postmodern flavor—surely it wasn't Dick Anderson, the septuagenarian Maine woodsman. Upon meeting Wylezol, I understood.

Before coming to Iceland, Wylezol had visited Greenland, which he assured me would be unlike any other section of the trail. Since there was no physical trail, hikers would be obliged to carry a map, compass, and (ideally) a GPS; since there were wide, frigid river crossings, they were advised to carry a small inflatable raft and telescoping paddle; and since there were polar bears, they were encouraged to carry a rifle.

That summer, René Kristensen, the director of the trail's Greenland chapter, and five native Greenlandic boys flew to Maine to see the beginning of the trail. The trip was an example of exactly the kind of cross-cultural interaction Anderson was hoping to foster. Though they were walking over roughly the same rocks as back home, the flora and fauna were totally alien; the only place they felt at home,

Kristensen said, was the barren summit of Katahdin. One night, a brief thunderstorm rolled through. The boys, for whom the Northern Lights are as common as clouds, were mesmerized. ("I've been in north Greenland for twelve years, and I have never, ever seen lightning," Kristensen said. "People don't know what it is.")

Outside the ambassador's house, I sat down on a couch next to Walter Anderson, the geologist. I asked him how Iceland fit into the larger picture of the Appalachian geology. "Oh, there are no Appalachian rocks in Iceland," he replied, shrugging. "In that way, it is a little bit artificial. But the reason the trail is coming here is because this lies directly on the Mid-Atlantic rift, where the two plates are separating. The rift is the connection. So, you see, it's part of the geologic story."

A few days later, the IAT committee members would all drive out to see the rift. When we stepped off of the tour bus at the Thingvellir National Park, we found a landscape swept bare of trees. Erupting through the grassy fields were rows of steep, flat-topped stone dusted with a sugary green. The sun dipped behind a cloud, and we felt a faint, black presentiment of winter.

The park's warden, Ólafur Örn Haraldsson, welcomed us with a short speech. Iceland, we were told, was born of division. Where the tectonic plates of Eurasia and North America pried apart, lava poured forth into the oceans, creating a submarine mountain chain. In one spot, an unusually active volcano continued to plume and pile up, rising tens of thousands of feet through the water column until it burst, black and steaming, above the waves. On the uppermost reaches of that peak, the nation of Iceland was built.

The two continental plates continued to widen at a rate of a few millimeters annually over the past ten thousand years, which, Haraldsson explained, had opened up the rift valley below. We wound our way down into the rift along a gravel path, the jumbled stone walls widening the deeper we went. Dick Anderson, gray-haired, slightly stooped in his blue rain shell, was lit up with childlike wonder.

"Oh my God, Don," he said at one point, gesturing toward a bird with a long orange beak. "What is that thing? What *is* that?"

"An oystercatcher," Hudson said.

"Wow," Anderson said. "They have *oystercatchers* here?"

As we moved deeper into the rift, he pointed to the rock wall and asked Walter Anderson, "Is this all lava?"

"All lava," Walter said.

"But some of it's different colors . . ."

"Well, that's bird shit."

At the bottom of the rift, we all posed for a picture by joining hands and stretching from one side of the rift to the other. The symbolism was clear: Wylezol announced that the Atlantic had once divided us, but now "the IAT is bringing us back together."

That sentiment also proved to be the refrain of the general assembly meeting, which had been held in a conference room at the office of the Iceland Touring Association the day before. Over the course of six hours, as committee members from around the world gave their presentations, an overhead projector cycled through photos and videos depicting the landscapes of Maine, New Brunswick, Newfoundland, Iceland, Norway, Sweden, Denmark, England, Scotland, Northern Ireland, Ireland, and Spain. When Don Hudson got up to speak, he pointed out that one of the original goals of the Appalachian Trail was to bring like-minded communities together. "Today, Benton MacKaye must be smiling on top of some ethereal mountain," he said.

The presentations went smoothly, but as the day wore on, people began, tentatively at first, to raise objections. First, a woman from Denmark asked whether countries where French was not the official language could drop the "SIA" (Sentier International des Appalaches) in the IAT/SIA logo. "Is it possible to make it more . . . *simple* in the future?" she asked.

Wylezol explained that the logo had been added to appease the Quebecois, but mentioned that it also worked in France and Spain (where the s could stand for "*sendero*"). "People may not understand why it's there," he said. "But it's like any logo, whether it be Mercedes Benz or whatever. It's just an image."

Don Hudson, who had invented the logo a decade earlier at his kitchen table, was less protective of it. "All these things are temporary and mutable, particularly when you think of geologic time," he said.

In the end, it was loosely agreed that trail clubs could put whichever three letters they liked on their blazes, so long as they kept the general shape and color scheme.

A woman from the Faroe Islands voiced her concern that hikers would become lost on the Faroese trails, which were marked only with cairns, and had always been navigated using only traditional knowledge. There was a brief, inconclusive discussion of how technology like GPS and QR codes could be incorporated into the trail. Then Startzell, the former director of the ATC, raised his hand and asked what the guidelines were for deciding how closely the trail must adhere to the geology of the Appalachians. "For example, if somebody proposed a section of the trail in Barcelona, which is I believe pretty well removed from the Appalachians, would we say, 'That's great, but that's not really part of what we're striving for'? Or is this kind of a come-one-come-all kind of thing?"

Wylezol replied that, "to the greatest extent possible," the trail should strive to adhere to the Appalachian geology. However, there were instances where that was not feasible. In Greenland, for example, the Appalachian chain ran up the east coast, but because the east was too difficult to access and navigate, the trail ran along the island's western edge. "Conceivably we could have a trail in Western Sahara," Wylezol added. "But right now I don't think anyone expects to go there and return with their two arms attached. So we have to be flexible."

Next, Wylezol asked those attending how linear they thought the trail should be. Before opening the topic up for discussion, Wylezol gave his opinion. His stance was decidedly nonlinear: He referred to the section he maintained in Newfoundland not as a trail but as a "route." Wylezol urged the attendants to keep in mind that "thru-hikers—the rock stars of hiking—as important as they are as a symbol and so on, are very, very few in the overall scheme of things." Most people would hike sections of the trail in (at most) one- or two-week stints, and for them, linearity was not such a pressing concern. In parts of the UK, the trail was split into spur trails in order to accommodate the most scenic areas of England, Scotland, and Ireland. "We don't want to abandon that just for some theoretical position that we have to be linear," he said.

Don Hudson again stressed the fact that connectivity, not linearity, was the trail's defining ethos. However, he admitted, strangers often had trouble wrapping their minds around this concept. When the trail had been extended to Newfoundland, people had said to him, "You can't walk to Newfoundland. How can that be part of the International Appalachian Trail?" His answer was that the trail existed "so long as we can describe to people how to get around the network."

A woman from Ireland excitedly called out that perhaps the trail could be renamed the International Appalachian Trail Network. Hudson mentioned that they had initially called it the International Appalachian Trails, plural. "I don't know why we dropped it," he said.

Finally, Wylezol asked the audience if anyone would like to speak in defense of strict linearity.

No one did.

+

To a great extent, the IAT was the offspring, and embodiment, of a very recent invention: that global reticulum of cables and code we call the Internet. The language and ethos of the Internet—a decentralized network of networks, designed to connect far-flung but like-minded people—infused (and facilitated) the trail committee's every discussion.

Fittingly enough, from its embryonic days, the Internet has always been an expansion upon one of the functions of trails (and, later, roads)—transmitting information quickly across long distances. Before roughly the nineteenth century, roads and pathways were the primary conduits of information in most countries. "Nobody lived more than a couple feet from the road until you got newspapers and telegraphs and internal combustion," Tom Magnuson told me. "In the age of muscle power, people lived as close as they could to the road, because that was the Internet. Every bit of your information came down that road."

With the advent of the telegraph, the dual function of trails (transporting matter; transmitting information) split. Matter followed one route, rolling along roads and railways (and water routes and air routes), whereas information was carried along wires, where it could travel far faster. And like trail networks, these wire networks created new ways of structuring the information they transmitted. As we saw in previous chapters, organisms as simple as slime molds use trails to externalize and organize information (*food is here; food is not there*), and indigenous human communities around the world have long used trails to make sense of landscapes, give shape to their stories (*first this happened here, then that happened there*), and to link together places of special (medicinal, spiritual, historical) interest. The dual invention of the computer and the Internet served as the latest breakthrough in our millennia-long search for better ways to transmit, store, sort, and process information.

In 1945 a prominent engineer named Vannevar Bush presciently anticipated the advent of the modern networked computer. That summer he published an essay in *The Atlantic Monthly* in which he envisioned a machine called the "memex" (memory+index). It would consist of a desk, two monitors, and a library of texts imprinted on microfiche—in addition to a great deal of technology that had yet to be invented, like touch-screens and modifiable print. Theoretically, wrote Bush, a memex user would be able to scroll through a series of linked documents while inserting his own links, comments, and edits. Looking back on this essay seventy years later, his device resembles nothing so much as a steampunk rendering of Wikipedia.

As a research scientist, Bush was acutely aware of the ever-amassing ocean of texts that our culture generated. Though he tended to blame science for this superabundance, in truth, the problem had been worsening ever since the invention of writing. The technology of written language allowed people to store information externally, which shifted us away from our reliance on oral storytelling and landscape-based memory. The advantage of this shift, as Sequoyah and his fellow Cherokees learned, was that the information did not noticeably decay when its author died, and it could be easily transmitted. The downside (aside from our eventual alienation from the land) was that texts began to amass more quickly than any one person could read them. The fear of information overload had been felt since at least the times of ancient Rome. The piling up of written information accelerated with the invention of the printing press, which prompted Renaissance scholars to invent organizing structures like indexes and tables of contents. Even then, scholars felt a desire to be able to link together texts into new forms. One crude eighteenth-century progenitor of the memex was a device called a "note closet," in which strips of cut-out text could be attached to slats and hung from hooks under various headings: information, made flexible.

In the emerging technology of microfilm, Bush saw the potential for information to be radically condensed and rearranged, allowing the *Encyclopaedia Britannica* to be "reduced to the volume of a matchbox." (Before the advent of digital technology, Bush had no way of knowing that, thanks to microchips, the *Encyclopaedia Britannica* would soon be shrunk to the size of a pinhead.) However, a desk full of a million tiny books would not on its own solve the problem of information overload; if anything, it would exacerbate it. To remedy this problem, Bush envisioned that the texts could be strung together into "associative trails." Largely, this task would fall to hardy souls "who find delight in the task of establishing useful trails through the enormous mass of the common record." It would be the job of these "trail blazers" to wade through the mass of information and connect them thematically, and then to share their trails, like guidebooks.

He gave the following example:

> The owner of the memex, let us say, is interested in the origin and properties of the bow and arrow. Specifically he is studying why the short Turkish bow was apparently superior to the English long bow in the skirmishes of the Crusades. He has dozens of possibly pertinent books and articles in his memex. First he runs through an encyclopedia, finds an interesting but sketchy article, leaves it projected. Next, in a history, he finds another pertinent item, and ties the two together. Thus he goes, building a trail of many items. Occasionally he inserts a comment of his own, either linking it into the main trail or joining it by a side trail to a particular item. When it becomes evident that the elastic properties of available materials had a great deal to do with the bow, he branches off on a side trail which takes him through textbooks on elasticity and tables of physical constants. He inserts a page of longhand analysis of his own. Thus he builds a trail of his interest through the maze of materials available to him.

In this scenario, Bush imagined that the scholar might later have a conversation with a friend who mentions that he is interested in the ways certain people resist innovation. Remembering the case of the Turkish bow and the English archers who failed to adopt it, the scholar could summon his trail, copy it, and give it to his friend, "there to be linked into the more general trail."

Bush's key insight was to realize that computers needed to evolve to fit the contours of the human brain. At the time, the prevailing mode of organizing information (whether in a filing cabinet or a library) was rigidly categorical and hierarchical. For example, to find a copy of Borges's *Ficciones*, one would begin by going to a library, traveling to the floor dedicated to literature, then to the section for Spanish-language literature, then to the row dedicated to authors whose names start with *B*, and so on. "Having found one item, [in order to find the next item] one has to emerge from the system and re-enter on a new path," Bush wrote. "The human mind does not work that way." According to Bush, thoughts are not grouped into categories; they are connected via "trails of association." By 1945 this was already a somewhat familiar notion, most famously elucidated in William James's *The Principles of Psychology*, in which he introduced the concept of "stream of consciousness." Bush believed that connecting the corpus of written texts in an associative manner would allow the human mind to make the most of what had always been writing's strength (and the brain's weakness): permanence of memory. "The personal machine," he wrote, will deliver "a new form of inheritance, not merely of genes, but of intimate thought processes. The son will inherit from his father the trails his father followed as his thoughts matured, with his father's comments and criticisms along the way. The son will select those that are fruitful, exchange with his colleagues and further refine for the next generation." Every step in the research process would be preserved: the stream of consciousness, he believed, could finally be frozen, extracted, and handed down over time.

Bush's essay was deeply influential on later generations of computer engineers, from Douglas Engelbart (an early visionary of personal computing) to Ted Nelson (the inventor of hypertext). The following decades saw the rise of two parallel but largely independent technologies: the invention of the personal computer, and the development of the Internet, which began by linking together various extant academic and military computer networks. The two paths fully converged with the invention of the World Wide Web and HyperText

Markup Language by Tim Berners-Lee. The pairing allowed people around the world to communicate and share information through a meta-network of personal computers, forming what Berners-Lee called "a single information space." Text, that hidebound technology, tendriled out into hypertext: documents were strung together into trails through hyperlinks; the trails sprouted side trails, which could loop back to the start; and a textual network emerged. The biographer Walter Isaacson has described this historic breakthrough as Bush's memex "writ global."

Like Bush, Berners-Lee also believed that texts should be malleable, so that readers could edit and improve texts as necessary. However, as the first successful web browsers, like Mosaic, began rolling out, Berners-Lee found, to his dismay, that they were composed of fixed columns of text surrounded by dazzling images, more like a magazine spread than a chalkboard—and, thus, more like a highway than a trail.

The Web has since sprawled, sending threads out across the globe, into each of our homes, our pockets, and almost inevitably, one day into our skulls. The current estimate of total web pages is almost fifty billion; if it were all bound into a single book, that tome would weigh over a billion pounds and would stand twice as tall as Mount Katahdin.

In recent years, people have begun to realize that the Web, which was designed as a tool to manage information overload, has ironically worsened it. A single trail reduces complexity and eases travel, but connect a thousand trails, and suddenly you have a maze that requires its own guide. Likewise, the Internet is a network of trails so vast it has become its own wilderness, "an uncharted, almost feral territory where you can genuinely get lost," wrote Kevin Kelly. "Its boundaries are unknown, unknowable, its mysteries uncountable. The bramble of intertwined ideas, links, documents, and images creates an otherness as thick as a jungle."

In the beginning, there was chaos, blank fields. Out of them, meaning emerged: first one trail, then another. Then the trails branched and webbed together, until they reached a density and complexity that again resembled (but was not quite) chaos. And so the wheel turned over. Benton MacKaye put it succinctly: "Mankind," he wrote, "has cleared the jungle and replaced it with a labyrinth." In this maze, a higher order of path making emerges—written guides, signposts, maps—which are then linked together and require yet higher orders of exegetical path making: written guides to the maps, and then guides to the map-guides, guides to the map-guide guides, and so on. (At first glance this notion sounds somewhat absurd, but I was recently amused to run across a medical text entitled *Guide to the Guides: Evaluator's Resource Algorithm to the AMA Guides to the Evaluation of Permanent Impairment, Fifth Ed.*) On each successive level of path making, knowledge is accrued and the world becomes easier to navigate, but new paths must constantly be marked out to simplify the vast wilderness of older paths into something humans can manage.

The function of the IAT, I realized, was yet another of these paths: it layered a higher order of guidance over the existing transportation network (which was in turn layered upon the older footpath network). And yet, in its desire to visit every place and connect everyone, the IAT was in danger of sprawling into yet another network, yet another wilderness.

+

Last on the schedule that day in Reykjavik was printed my name, and beside it, the word *Morocco*. Rising to the podium, I tried to tamp down the audience's expectations as I loaded up the photo slideshow of the scouting trip I had taken to the trail's end. The committee members were expecting camels and deserts—Wylezol had said as much at least twice in the hours preceding—whereas my photos mostly contained stony hillsides, olive orchards, old televisions, and flat red fields haunted by dogs.

Many months earlier, back in Portland, when I had asked Dick Anderson where the trail would end, he told me they had agreed on a town called Taroudant in the Anti-Atlas Mountains. He said the word *Taroudant* slowly, with a faint smile, as if it held some exotic magic. He told me a former Peace Corps volunteer had mapped out the route from Marrakech to there. However, when I wrote to the volunteer, he told me that Anderson was mistaken. The final stretch of trail had yet to be mapped. If I wanted to see the trail's end I had a choice to make: I could wait for someone else to map it out, and write about it while peering over her shoulder (in my usual, wraithlike way), or I could go there and map the thing myself.

I wrote to Anderson and told him to expect a mapped route to Taroudant by the end of the spring.

Initially, I imagined I would travel to Morocco and, map in hand, explore the desert wilderness between Marrakech and Taroudant. However, I soon scrapped that plan as fantasy. Between Marrakech and Taroudant lay not a wilderness, but a vast swath of hillside farms, pastures, and mountain hamlets. Since I spoke almost no French and absolutely no Berber, I would have been unable to converse with most of the people I met there. (Only fifteen percent of all Moroccans speak English, and even less in the remote mountains.) The topographic maps I had ordered of Taroudant, care of the Russian military, revealed a network of hundreds of spidery trails, marked with faint dotted lines, on which one could easily become lost. I knew I would need guidance through this network, so with Anderson's help I hired a local guide named Latifa Asselouf. She promised to arrange everything, including meals and lodging.

When I arrived at the Marrakech airport, a driver was waiting for me with a sign. In lieu of a hello, he had handed me his cell phone. It was Asselouf.

"Hello, Robert? This is Latifa. The driver will take you to my home now."

"Great," I said. "Thanks."

She hung up.

In an attempt to be friendly, I tried to ask the driver his name.

"Je ne parle pas l'Anglais," he replied, apologetically.

"D'accord," I said. I asked him again, this time in my halting French. He handed his phone to me. It was Latifa again.

"Hello, Robert? The driver, he does not speak English."

"Okay, thanks." I said.

He led me to a battered white Mercedes. As our car slipped loose of the pink city of Marrakech, I looked out the window and took note of things—a horse cart hauling bags of grain, a herd of goats parting fluidly around our car, two women riding on a motorbike with a child sandwiched between them—and then I took note of the fact that I was only taking note of things that seemed "Moroccan," as opposed to the things our two countries had in common: the garish advertisements, the electric wires zigging up the valleys, the paved roads swarming with cars, the red-and-white cell towers standing like the skeletons of decommissioned spaceships.

As we entered the town of Amizmiz, the air grew cool. Asselouf was standing outside her front door, smiling broadly and wiping her hands on a dish towel. She had the body of a walker, thin and long-limbed. Unlike her light-skinned neighbors, her skin was deeply tanned—a mark of her family's Saharan ancestry. Her "crazy hair," as she called it, was tied back with a violet headscarf.

She ushered me into her living room and set down a clay *tagine* filled with stewed lamb and prunes. A man wearing a red windbreaker and a small black watch cap walked in, sat down beside me, and shook my hand. He had a thin face, a prominent nose, and a small black mustache. His name, Asselouf informed me, was Mohamed Ait Hammou. He had been hired to serve as our pathfinder, while Asselouf would handle the logistics, accommodations, and my endless barrage of questions. He did not speak English, so, while Asselouf was busy in the kitchen, we ate in silence.

After lunch, we piled our things into a microbus and drove to a town about an hour up the road. As we rode, Asselouf pointed out the flora that grew on the terraced hillsides: groves of gray walnut trees and pink peach blossoms; gardens of mint, thyme, and tulips. The walls of the buildings were constructed out of flat stones, which had been piled like loaves of bread in a shop window, a snug jumble. From far off, the villages disappeared, camouflaged into the rocky hillsides, with the exception of the white mosques and perhaps a newer, concrete house painted Marrakech pink. The local men and boys mostly wore tan or brown *djellabas*, long robes with peaked hoods, which gave them the air of Franciscan monks.

These roads—unlike roads built in the United States and most of Europe—were mostly made by voluntary communal labor; villages came together to build and maintain them. A few days later, we would pass a group of eight smiling men who were constructing a crib wall to reinforce the road that connected their two villages. A ninth man crouched over a fire nearby, making tea.

At a certain point, the van stopped and Asselouf signaled for us to get out. We grabbed our backpacks from the roof and began walking down the road. It was now cold, and the sky was growing dim. We walked for about an hour. Then, as she did every night that week, Asselouf approached the nearest village and began asking around if any of the local residents would give us dinner and a place to sleep on their floor (for a reasonable fee). That night, we stayed in a small home overlooking a vast, vapor-capped valley. Before dinner, the men all sat together watching television with our legs under a single blanket in the living room.

The newscaster on TV was dressed in a black blouse; her hair fell to her shoulders in dark curls; she sat with her hands folded at a desk, with electronic graphics flashing in the background. In other words, she looked like a newscaster on any channel in America, except that the words she spoke meant nothing to me. Sitting there in my synthetic clothes, pantomiming comprehension, I felt oddly at home and yet profoundly out of place.

The Germans have a word for precisely this feeling: *unheimlich* (literally, "un-homelike"). According to the theorist Nicholas Royle, the *unheimlich* (often translated as "the uncanny") is defined as "a peculiar commingling of the familiar and unfamiliar." We are comfortable with the familiar, and we are comfortable with the wholly unfamiliar (which we perceive as exotic), but when the two are combined, we begin to feel unstable. The result, Royle writes, is "the experience of oneself *as* a foreign body."

+

The next morning, the three of us hiked up and over a ten-thousand-foot pass, following sheep trails. Before we departed, Asselouf had hired a local mule driver to carry our backpacks. As we neared the top of the pass, I learned why: one needed to move quickly through this cold and otherworldly place. Mist blew up from the valley below, cottoning us in. The downhill sides of every plant were shrink-wrapped in layers of ice: the tall grass became icy white feathers; shrubs resembled coral reefs. The mule driver blew into his hands, his curly hair collecting little nerds of ice. At one point he got spooked and tried to abandon us, pulling our packs off the mule and setting off in the wrong direction. In the end, Asselouf had to double his wage to lure him back.

On the other side of the ridge, in the lee of the wind, Hammou knocked some dry roots out of the ground with a sharp rock, built a small stone fire pit, and kindled a fire. Asselouf made grilled lamb kofta, with olives and fresh pita. It was, without question, the most delicious lunch I had ever eaten on the trail.

Afterward, we headed down the left-hand side of a rocky gorge.

During our descent, Hammou took a shortcut that required us to scramble down a scree slope, which neither Asselouf nor the mule driver liked very much. Hammou, meanwhile, was as nimble as a goat. Near the bottom of the pass, he stopped and waited for us while pecking at his cell phone.

Over the coming days I would learn that Hammou had a tween-like obsession with his cell phone; he could walk for miles without looking up from it once. I asked Asselouf what he was always doing on it. She said he was probably texting one of his many sweethearts; he seemed to have one in each village (in addition to the wife he had at home). Later in the trip, he would often find an excuse to reroute us through the towns where they lived, so he could spend a few minutes (or, in one case, a few hours) flirting with them in their homes, while Asselouf and I waited outside. Afterward, Asselouf would chide him for wasting our time and being unfaithful to his wife. He would respond by scowling at his phone.

Hammou seemed to particularly resent the fact that his boss was a woman. Whenever one of his friends called to ask how Asselouf and I were faring, he would joke that Asselouf was crying. Once, Asselouf told me, he turned to her and asked, "Why don't you go home and raise babies like a normal woman?" Asselouf shrugged off these provocations. She'd heard worse.

From the outset, Asselouf's chosen line of work had been a struggle. Back when she was in her early twenties, she had told her mother she wanted to take a course in mountain guiding, but her mother forbade her. When Asselouf persisted, her mother slapped her face. Asselouf went anyway. She was now thirty-nine years old. She lived in the same home where she and her seven siblings had grown up. They all eventually moved out, leaving her to take care of her ailing mother, alone. Currently, she was one of only two female guides she knew of in all of Morocco.

In the Berber highlands, she stuck out. She dressed unconventionally, in gray yoga pants, a knee-length merino sweater, and a black rain shell. When we passed through villages, she was regarded as the oddity, not me. The children gathered behind her, speculating in whispers whether she was from France or America. When she turned on her heels and told them that she was a Berber too, they burst into fits of nervous laughter.

The three of us made for odd travel companions; though we were walking the same path, we each had different goals. It seemed Hammou simply wanted to get us to Taroudant as easily as possible, get paid, and go home. Asselouf was trying to expose me to the local culture and natural beauty, while also heading off any potential disaster. And I was here, above all, to chart a potential route for the IAT. Asselouf often tried to communicate my rationale to Hammou, but it was clearly a struggle. *He's here to map out a very long hiking trail that will one day stretch from North America and Europe to Morocco,* I imagined her saying in Berber. I could tell by his reaction he found the whole idea a bit preposterous.

Indeed, I was beginning to have doubts of my own. Aside from Hammou and Asselouf, my interactions with Moroccans had mostly been fleeting—a smile, a wave, but little more. I realized hiking might be a deeply inappropriate means of connecting disparate cultures. To truly connect to the people living here, one would need to stop for a year (or ten), set down roots, and learn the language. Hiking is about movement, a continual sliding over the surface of things. What meaningful cross-cultural communication could possibly come of it?

On our second night, we stayed in a home where everything smelled of fresh paint; the walls of the main entry hall had been painted robin's egg blue in preparation for an upcoming wedding. We sat in the kitchen as the matriarch of the household poured out glasses of milky Berber tea, redolent of thyme, and ordered around her five daughters and four granddaughters. Later, the patriarch ambled in, a ninety-two-year-old former judge with ears like giant moth wings. In a show of deference, everyone offered him their seat. Asselouf wisely intuited that the old man would like to rest his back against a hard surface, so she got up and moved a stool against the wall, where he gladly eased himself down. (Later, Hammou complained that Asselouf concerned herself too much with other people's feelings. "That," she replied, "is why I am the best guide in all of Morocco.")

Once we had exchanged pleasantries, Asselouf began peppering the judge with questions about the local topography. As part of his job, he had traveled extensively throughout the mountainous countryside, settling cases and arbitrating disputes, so he knew the name of every village and the most efficient route over every mountain. (He praised the French for widening trails and cutting new roads, if nothing else.) After a while he pointed to me and asked Asselouf something. I watched her explain, over the course of many minutes, using vivid hand gestures, the story of the breakup of Pangaea, the cleaving of the Appalachian Mountains, and the proposed trail that would link them all together. I turned to watch their faces, expecting to see bewilderment or mockery. The old man nodded his head slowly and said something. "He says that was the will of Allah," Asselouf said. "He says that long ago we all came from the same place, but just this"—she pinched her cheek—"became different. Beneath, bones, blood, it's all the same. You understand what I am saying?"

+

We hiked on, over snowy mountain passes, along dirt roads and sheep trails. We hired yet more mule drivers, including one young man who routinely beat his mule with a wooden club; every few minutes the mule would retaliate by lifting its tail and releasing a pneumatic stream of gas into his face. The hills varied between gray, taupe, and blood-black. When the continents were joined in the Pangaea supercontinent, parts of Morocco would have nestled up against northern Maine. But I had trouble seeing the kinship between these sandy mountains and my green-backed, granite-spined Appalachians. (In fact, I would later learn, we were still in the High Atlas, which was a much younger formation, geologically speaking. Only later would we reach the Appalachian geology of the Anti-Atlas.)

Two days later, we caught a microbus and skipped about fifty kilometers of the trail, so we could make it to Taroudant in time for me to catch my flight. Asselouf was careful to note the names of all the villages we would have passed through, so I could mark them down on my map later. On the bus, a young man with a round face offered to let us stay at his house. Asselouf eyed him suspiciously.

"Is your house clean?" she asked.

"You'll see," he said.

"Do you wash your dishes?"

"You'll see," he said.

Before she could go on, he preempted her. "Please, stop asking questions. Do you want to stay or not?"

When we finally arrived at his house—which lay at the end of a long road cutting through farms and fruit orchards—he turned to Asselouf and said, "Now you see. Everything is dirty."

Asselouf sighed. Without another word, she began rinsing out the unwashed tea glasses and sweeping the kitchen floors.

The house was little more than a concrete shanty. The layout was bisected. Two of its four rooms were for humans: a kitchen, with a dirt floor and benches made of cinder blocks and wood boards; and a bedroom, where blankets covered a hardwood pallet. The remaining two rooms were dedicated to housing a milk cow. In this small, disheveled space lived four cousins. The three oldest—ages twenty-three, nineteen, and seventeen—worked on nearby watermelon farms and orange orchards. The youngest of them, who was only twelve, worked as a shepherd.

To illuminate the room, one of the boys lit a large, swan-necked propane lantern, which resembled something from a chemistry lab. We sat down on the bedroom floor, and Asselouf served a dinner of lamb stew in a tagine. The boys, who were used to simple bachelor fare like lentils and rice, devoured the stew, peeling off hunks of flatbread to sop up the juices. They were shy at first, especially the youngest, who hid behind his brothers and peered at us with suspicion. Before long, though, Asselouf was playing Wendy among the Lost Boys; she taught them phrases in Arabic and French, and teased them for their messiness. By the end of the night they were begging her to move in with them.

After dinner, the oldest of the boys, Abdul Wahid, challenged me to a game of checkers. The checkerboard was a piece of plywood where the squares had been drawn in by hand. For the pieces, we used small rocks he gathered from the yard. Checkers is a very old game—a three-thousand-year-old checkerboard was found in the biblical city of Ur—but its modern form was shaped by the French, who most likely introduced it to the Berbers. The French rules, which Abdul Wahid played by, were slightly different from those I had grown up with. (For example, when a piece was "kinged," it could travel diagonally across the board, like a bishop in chess).

When our game was finished, the boys started playing with Latifa's camera. The two younger boys took turns posing while wearing my glasses and pretending to read my copy of Waiting for the Barbarians. Then Abdul Wahid posed with my iPhone held to his ear. At first I thought they were poking fun at me, but as the boys gathered around Asselouf's camera to inspect the pictures of themselves, I realized it was a kind of play-acting. They were imagining themselves into a different life.

"Real travel," wrote Robyn Davidson, "would be to see the world, for even an instant, with another's eyes." However, I was discovering that this process works both ways: a journey is never simply the act of gaining a new perspective, but also the experience of being *newly seen*. Again, I was overwhelmed by a feeling of the *unheimlich*, but this time, surrounded by kind, curious faces, the feeling was warmer, more expansive. Boundaries were dissolving.

As the night wound down we cleared away the dishes and lay down together—Hammou, me, and Asselouf—on the wooden pallet in the living room. In the kitchen, the Lost Boys were bedded down on blue plastic bags of grain. A radio warbled gently in the background as they whispered themselves to sleep.

+

In the morning, we set off for Taroudant. The land that lay ahead was a hard clay pan between two mountain ranges, known as the valley of Souss. We passed by watermelon farms, citrus orchards, wheat fields, and groves of Argan trees. Heat wavered up from the earth and was dispersed by the wind; it felt like walking across the scorched bottom of a *tagine*. At one point, we were chased by three scrawny, vicious dogs, which we had to fend off with rocks. When we asked their owner why he didn't call off his dogs, he smiled and replied that we shouldn't be walking across his land if we didn't want to get chased.

In the mountains to the north of Taroudant, I had hoped to find a peak that would provide a suitable terminus to the trail, a Moroccan Katahdin. However, we were running behind schedule, and, without telling either me or Asselouf, Hammou began altering the plans to catch us up. It wasn't until around midday that I realized we weren't taking an indirect route to the mountains, as I had assumed, but were instead cutting a long hypotenuse across the flat farmlands directly to Taroudant. I turned to Asselouf and asked her what was going on. She asked Hammou, and then explained that he had taken this shortcut because it was faster and "closer to modern things."

It was too late to change our route; we would never reach the mountains and make it back to Taroudant in time. I was disappointed, but not surprised. I had watched Asselouf attempt to explain the trip's rationale to Hammou on multiple occasions, always to his utter bafflement. He seemed to have no sense of why anyone would voluntarily choose to hike through mountains. Over the course of our hike, he had often taken drastic shortcuts through less-than-scenic areas—on one memorable occasion, to save a few minutes, he had led us down a gulch snowed over with balled-up plastic diapers. Now, in one last shortcut, he had lopped off the journey's last, most important segment.

My flight home was scheduled for the following day. While I had done enough legwork to give the IAT board members a sense of what the trail might look like, I hadn't done nearly enough to recommend a route. Asselouf and I talked through our options, and we agreed that she would have to complete that task on her own. Which, in retrospect, was how it should have always been. These were her lands to map, her story to tell.

Weeks later, she wrote to inform me that she had returned to the High Atlas above Taroudant, where she passed through forests of oak trees, slept in goat huts, and eventually climbed to the lofty mountain hamlet of Imoulas, from which one could reach the towering twin peaks of Jbel Tinergwet and Jbel Awlim. Strictly speaking, neither bears Appalachian geology, but photos reveal that both are eerily reminiscent of Katahdin—which would in some odd way make for an even more fitting conclusion to this long, strange, puzzling epic of a trail.

+

Back home, I found myself wondering about the ultimate endpoint of the IAT. The last time I checked, the trail committee was still in the process of setting up a local chapter of the trail in Morocco and consulting with guides, including Asselouf, to decide on a terminus.

If history is any indication, whichever mountain the Moroccan chapter chooses will likely prove an unstable endpoint—more of an ellipsis than a full stop. The impetus of a long trail is to grow ever longer. The architects of the AT once shifted its terminus from Mount Oglethorpe to Springer Mountain, and then from Mount Washington to Katahdin. Then Dick Anderson extended the trail from Katahdin up through Canada, and then again, over the Atlantic and down to

Morocco. But the world's longest trail could feasibly grow even longer. Technically, the Appalachian range continues south, far beyond Taroudant, into the disputed nation of Western Sahara. Likewise, on the other end of the trail, remnants of the Appalachian range technically continue beyond Georgia, through Alabama, all the way to the Wichita Mountains of Missouri and the Ouachita Mountains of Oklahoma. Delegations from both states had been lobbying the IAT for an extension. "It would be a hell of a walk," Walter Anderson, the geologist, told me, "but their scientific rationale is perfectly legitimate."

Curious as to how far the trail could ultimately stretch, I called up a few geologists. One told me that his research suggests there is a pocket of Appalachian rocks in southern Mexico, which was marooned by the opening of the gulf as the continents drifted. Another told me she hadn't heard about the Mexican Appalachians, but she had heard there might be Appalachian remnants in Costa Rica. Yet another geologist could not vouch for the Mexico and Costa Rica theories, but had reason to believe there might be traces of the Appalachians as far south as Argentina.

When I next talked with Dick Anderson, I mentioned what these geologists had told me. I half expected him to grow defensive, but he seemed to find the notion delightful. "The way this project has worked is that it's expanded as people wanted to expand it. We don't have any big campaign to expand it," he said. "But we're willing to go wherever that original principle leads us."

Before my hike through the Atlas Mountains, that original principle sounded noble and ambitious: to trace the remains of an ancient, scattered mountain range; to grapple with the immensity of geologic time; to blur political boundaries; and to connect distant people and places. But when I arrived in Morocco it had suddenly begun to seem wildly idealistic. Unlike in America, the Moroccan leg of the trail would not pass through empty parklands; much of the trail's length would be inhabited. I wondered what would happen when thru-hikers began arriving from Georgia. Would the local people generously welcome them, as they had me, or would they grow irritated by the steady trickle of camera-wielding strangers? And what about the hikers? Would they have respect for the local people, or would they regard them—as hikers so often have throughout history—as pests befouling an otherwise pristine landscape?

I had begun to doubt whether mere physical connection—mere trails, mere highways, mere fiber optics—could bridge meaningful divides between people. In the era of the jet and the Internet, the world is in many ways more connected than it has ever been. But there is another meaning of connection that our networks don't capture, what we refer to when we say that we "have a connection with someone." The philosopher Max Scheler has called this intimate quality *fellow feeling*—a sense of deep, mutual understanding. He argued that this type of connection requires us to recognize that the minds of other people have "a reality equal to our own." This recognition in turn allows us to extend beyond the confines of our individual minds to more bonded, collective ways of thinking. "It is precisely in the act of fellow feeling," Scheler wrote, "that self-love, self-centered choice, solipsism, and egoism are first wholly overcome."

The problem facing the IAT's planners was that this kind of connection—bound up in the slow-shifting and still largely mysterious landscape of the human brain—cannot be accelerated at the same rapid rate as other forms of connection. We can travel at the speed of sound and transmit information at the speed of light, but deep human connection still cannot move faster than the (comparatively, lichenous) rate at which trust can grow.

This is the unexpected disconnect that a vastly interconnected landscape ends up creating. Connection without fellow feeling invariably leads to conflict; when two cultures are abruptly put in contact, the differences between the two groups often jump out in sharper relief than the similarities. For example, when Europeans first crossed the ocean and encountered Native Americans they became fixated on their differing religious and cultural values and overlooked their commonalities. The result was centuries of warfare, followed by an exploitative power imbalance that continues to this day. This same dynamic replayed itself countless times throughout the history of imperialism.

In recent decades, with the rise of globalization and mass communication, though the cultural differences between nations have greatly lessened, this sense of contrast has only grown more visible. Because far-off places now *feel* so close, and because it takes less work to make contact, we assume that the people elsewhere will share our way of seeing the world. When they *don't*, we often conclude that they are foolish or bad or irremediably strange. If we can be in direct contact with someone and still feel so distant, one starts to wonder, how can that distance ever be bridged?

On my hike through the Atlas, I had mulled over this question many times in regard to our pathfinder Hammou. I spent a week with him; we walked together, ate together, and slept side by side on a wooden pallet together. I often tried to converse with him, with Asselouf acting as our translator. But at the end of the trip I still felt no sense of fellowship with him. His approach to many aspects of his life—his derisive attitude toward Asselouf, his penchant for checking his text messages rather than admiring the mountains, his near-comical affinity for shortcuts—grated against my own.

On top of everything else, Hammou and I differed radically in our regard for the landscape. I had often felt that Hammou saw the Atlas Mountains as nothing more than impediments to our progress, and that he, like the New England farmers of yore, would gladly flatten them if he could. This kind of thinking in the United States had ultimately led to a deeply destructive mind-set, the most obvious manifestation of which was a string of notorious Appalachian mining operations involving the literal removal of mountaintops. But I had overlooked the possibility that Hammou, who had grown up among these mountains, might have a connection to them that was subtler but vastly more intimate than what I felt for my beloved Appalachians.

I came to this realization as the three of us sat in our shared hotel room in Taroudant that final night of our trip, trying to reconstruct our journey on the map. Asselouf turned to Hammou for help at one point, and I watched him recite the names of each town, mountain, and landmark we had passed that week, entirely from memory, with a faint, fond smile on his lips:

. . . Taddaret,
Akhferga,
Wawzrek,
Al-Khoms,
Toug-El-Hir,
Tazlida,
Tnin-Tgouga,
Tamsoult,
Tagmout,
Imamarn,
Tazoudout,
Talakjout,
Larba,
Tizi-N-Al Cadi . . .

At the time, I had known too little about where Hammou had come from—and too little about where *I* had come from—to fully grasp the nature of our disagreement. Wild landscapes inspire awe in Euro-Americans—the descendants of ruthless conquerors, raised on a continent rich with natural resources—because for generations we have used our wealth and technology to insulate ourselves from the land's harshest elements. The Berbers, meanwhile, having avoided the worst ravages of industrialism but having suffered the inequities of colonial rule, never rebounded into a romantic love of wild nature. "They don't see it as a recreational area," I was told by Michael Peyron, a visiting professor at Al Akhawayn University and an expert in Berber poetry. "They see it as a place where they live, they see it as a challenge, and now, of course, they see it as a place of earning money." Peyron added that many Berbers also visit mountaintops to make sacrificial offerings or to visit the tombs of Muslim saints. I recalled that Hammou had mentioned that when he was feeling especially distraught, he would climb a high peak so that the immensity of Allah's creation would help put his problems in perspective. There are many ways to love a landscape.

What Hammou and I were ultimately lacking, it seems, was not sufficient contact, but sufficient context. The rifts between two people can easily appear unbridgeable at first, like the void between two peaks. But when we peer deeper into the chasm—down through the complex strata of culture, technology, and happenstance—we often find a shared point from which one could start scaling either summit. As the old judge had said, "Long ago, we all came from the same place." From these common origins, humans have branched out across the planet, adapting in multifarious ways to the land and to one another, diverging and converging, disconnecting and reconnecting, growing foreign and getting reacquainted.

+

When I think back on that trip now, years later, there is a moment that stands out from the others. It was our last day, during the long march across the Souss valley toward Taroudant. We were following a wide dirt trail that ran through sparsely planted groves of fruit trees. I was walking behind Hammou, and Asselouf was walking behind me. I was lost in some dark thicket of my own thoughts, when Asselouf called out in surprise. Hammou and I stopped and turned around. She called out again, pointing off into the scraggly orchard. Our eyes followed her index finger to discover what the two of us had blindly walked past: a goat, perched impossibly in the upper branches of an Argan tree, some fifteen feet in the air, straining its lips toward the highest branch to reach a cluster of olive-shaped fruit.

Asselouf told me that goats all throughout the region had learned to climb these trees. Once the goats had digested the Argan fruit, farmers collected the seeds from their droppings, which they then pressed to extract the oil. They sold this oil for astronomical prices to foreign countries, where it was rumored to reverse the skin's natural aging process.

The goat nibbled at the hard green fruit. As we stood there gazing up at it, a slow joy welled in me. There was something familiar about the taut tendons of its neck, the nervous adjustments of its little hooves on the narrow branches. I suddenly felt a deep kinship with it—and with all the rest of us restless creatures, forever striving toward something just beyond our grasp.

I will never know precisely what Asselouf and Hammou were feeling at that moment, but when I looked over at them, I noticed they were smiling too. Asselouf took a photo of the goat in the tree. She promised to send me a copy. Then we shouldered our packs, turned our eyes back to the trail ahead, and began, once again, to walk.

EPILOGUE

WE MOVE through this world on paths laid down long before we are born. From our first breath, there is a vast array of structures already in place—"spiritual paths," "career paths," "philosophical paths," "artistic paths," "paths to wellness," "paths to virtue"—which our family, society, and species have provided for us. In all these cases, the word *path* is not applied haphazardly. Just like physical paths, these abstract paths both guide and constrain our actions—they lead us along a sequence of steps, progressing toward our desired ends. Without these paths, each of us would be forced to thrash our way through the wilderness of life, scrabbling for survival, repeating the same basic mistakes, and reinventing the same solutions.

There is a catch, however: How do we know *which* paths to choose? The essayist James Fitzjames Stephen vividly captured this dilemma: "We stand on a mountain pass in the midst of whirling snow and blinding mist, through which we get glimpses now and then of paths which may be deceptive. If we stand still we shall be frozen to death. If we take the wrong road we shall be dashed to pieces. We do not certainly know whether there is any right one. What must we do?"

Even a cursory reading of ancient philosophy reveals that it has never been easy to choose a path through life. But it is becoming ever more difficult. Rapid changes in technology, culture, education, politics, trade, and transportation have combined to allow people access to an array of lifestyles that was previously unthinkable. In the aggregate, this is a positive development, proof that our life's paths are evolving to meet our varied desires. But a side effect of this shift—halting, gradual, and unevenly distributed as it may be—is that life's options continue to abound until they overwhelm.

Take, for example, the commonplace question of what you are going to "do for a living." In the earliest days of humankind, there was likely just one answer: gather plants and scavenged meat, an activity all people participated in equally. Later, new specializations emerged: first, the invention of hunting, then of medicine, shamanism, arts, and agriculture. According to the Standard List of Professions, a five-thousand-year-old catalog of occupations from ancient Mesopotamia—ranked in descending order from king down to some as-yet-untranslatable, but surely unpleasant, job—there were 120 separate professions on offer. Today, it is estimated there are anywhere from twenty thousand to forty thousand distinct occupations in the United States.

Our selection of religious and philosophical traditions is scarcely less varied. Due to the difficulty of defining what constitutes a proper religion, estimates differ, but most agree it is easily in the thousands. And this is merely a tally of the organized religions; the number of personal belief systems—cobbled together in private, one piece at a time, like the car in the old Johnny Cash song—is impossible to quantify.

In the end, we are all existential pathfinders: We select among the paths life affords, and then, when those paths no longer work for us, we edit them and innovate as necessary. The tricky part is that while we are editing our trails, our trails are also editing us. I witnessed this phenomenon firsthand on the Appalachian Trail. The trail was modified with each step we hikers took, but ultimately, the trail steered our course. By following it, we streamlined to its conditions: we lost weight, shed possessions, and increased our pace week after week. The same rule applies to our life's pathways: collectively we shape them, but individually they shape us. So we must choose our paths wisely.

+

When I returned to New York City from the Appalachian Trail in 2009, the experience of thru-hiking lingered in my bones. The intricate machinery of my feet—the tarsals and phalanges, the cuboid and cuneiform bones, the ligaments and tendons, the muscles, arteries, and veins—ached for a month afterward. In the mornings I would rise from bed and hobble to the bathroom with cringing, nonagenarian steps.

Thru-hiking is metamorphosis: over five months, I had acquired a new name, a new body, a new set of priorities. By the trail's end, I was as trim and clear-headed as a wild animal. But back home, over a matter of months, I gradually regressed into something resembling my old self. First, I shaved my scraggly, Mansonesque beard, which had begun to draw nervous stares from strangers; then, a few weeks later, I cut my hair. The weight I had shed slowly filled back in, layer by later, as if I were being dipped in paraffin. I went back to living in a box full of possessions and spending my days staring at glowing screens. The path of least resistance, that old rut, drew me inexorably back in. As the architect Neil Leach has noted, "The city modifies its occupants, no less than the occupants modify the city."

I often thought about a very old poem I had once read, by a mountain hermit in ancient China named Han-shan.*

The Cold Mountain trail goes on and on:
The long gorge choked with scree and boulders,
The wide creek, the mist blurred grass.
The moss is slippery, though there's been no rain
The pine sings, but there's no wind.
Who can leap the world's ties
And sit with me among the white clouds?

Han-shan was raised in a thriving metropolis and groomed for life as an imperial envoy, but at age thirty, he left home and traveled a thousand miles east to a cave on the slopes of Cold Mountain, where he remained for the rest of his life, writing poems and "wandering completely free." Upon moving there, he took the mountain's name as his own: Han-shan is a trail name of sorts, which means Cold Mountain. His needs were few; his pillow was a "boulder," and his quilt was the "dark blue sky." What he desired from this new life, he wrote, was to lie down in a cold clear stream and wash out his ears.

Han-shan would eventually become one of China's most beloved poets, and a hero to seekers and vagabonds around the world. His poems often return to the dichotomy between the roads of town life, which he avoided, and narrow mountain trails, which he sought out. Buddhists and Taoists both, of course, had long employed the metaphor of the trail to describe their philosophies, but the Tao and the Dharma were portrayed as broad paths, for one and all. Han-

---

* This translation comes from Gary Snyder's *Cold Mountain Poems.* The other translations in this chapter have all been taken from Red Pine's *The Collected Songs of Cold Mountain.* Both are excellent; Snyder's is more lyrical, while Red Pine's is more academically precise and exhaustively complete.

shan broke from this tradition: He believed that a way of life could become too common, that a trail could be too crowded or too worn; he urged his readers to "leave the dusty rut behind" and seek out "paths of newly trampled grass." A millennium later and half a planet away, Thoreau would make this same metaphorical connection: "The surface of the earth is soft and impressible by the feet of men; and so with the paths which the mind travels," he wrote. "How worn and dusty, then, must be the highways of the world, how deep the ruts of tradition and conformity!"

What these men were describing, I realized, was a phenomenon of path-breaking common to all living things: One caterpillar finds a new leaf, and ten more follow its trail. By the time the eleventh arrives, the leaf has been chewed down to its skeleton, and so the eleventh caterpillar grows hungry and sets off in a new direction. The same principle applies to foraging ants and grazing herds, to fashion trends and stock markets, to traffic-clogged roads and eroded hiking paths. By striking off into the "darkest wild" of the Tiantai Mountains, Han-shan found a space free from suffocating conventions, where he could live a refreshingly stripped-down existence.

In the years following my thru-hike, I began to realize that the Appalachian Trail had been my Cold Mountain—a wild space defined by simplicity and freedom, relatively untouched by violence or greed, with a clear goal and few distractions. But unlike Han-shan, I had left it and returned to the metropolis. That other life haunted me.

+

It was possible, I gathered, to spend one's life doing little else but walking. Life on the trail being exceedingly cheap, a handful of full-time hikers have managed to live for years or even decades off meager savings and seasonal work. These wanderers reminded me of mendicant monks, slipping free of the gravitational pull of society to live plainly, outdoors.

Over the years a curious name kept popping up, in the oddest places, as an example of a perpetual hiker, perhaps *the* perpetual hiker. This man called himself Nimblewill Nomad. Following the completion of his first thru-hike of the International Appalachian Trail in 1998, the Nomad had reportedly given away all his money and taken to hiking long trails more or less full-time. He proudly referred to himself as "hiker trash," a modern update on the archetypal tramp. I'd heard he spent his winters living out of his pickup truck, sleeping in Walmart parking lots and national parks. As soon as the weather warmed, he walked.

His exploits tended to take on a mythic cast. I heard from more than one person that the Nomad had chosen to have all ten of his toenails surgically removed to avoid infections. Famously minimalist, he was said to never carry more than ten pounds on his back. People joked that his cook kit consisted of nothing more than a bent spoon and a cigarette lighter. Whenever possible, to avoid carrying food, he opted to eat at cheap roadside diners and gas stations.

This style of hiking was not universally admired. Lamar Marshall, who met the Nomad in 2001, told me he thought that hiking marathon distances to reach a restaurant each night "defeats the whole damn purpose of being in the woods." But the Nomad, it seemed, had long ago moved *beyond* the woods. I was fascinated by his refusal to respect the boundary we have erected between the human world and the natural one. Each day, he somehow carved out a graceful path through the many-chambered heart of the behemoth—wandering the postindustrial wilderness, from forest to forest, fryolator to fryolator, in what he called "a desperate search for peace."

Over fifteen years, he had hiked some thirty-four thousand miles. First he completed the so-called Triple Crown: the Appalachian Trail, the Pacific Crest Trail, and the Continental Divide Trail. Then he went on to complete all eleven National Scenic Trails, an achievement that has somewhat awkwardly been deemed the Undecuple

Crown, in 2013. He finished that hike atop Mount Monadnock, where he was congratulated by Dick Anderson and a host of other friends. Triumphant, fulfilled, and nearing his seventy-fifth birthday, he vowed to hang up his hiking shoes. Then, the next spring, he was back. He announced he would complete a grueling road walk from New Mexico to Florida, in order to complete a route he had named The Great American Loop, which connected the four farthest corners of the continental United States. This, he claimed, would be his last long hike.

Without having met him, I had no way of knowing whether the Nomad was merely a bitter misanthrope or, in the words of Jack Kerouac, "a new kind of American saint." I wanted to see what this lifestyle truly entailed, and what kind of man it shaped. So I wrote to him one afternoon to ask if I could join him for a few days on his final hike. After some delicate negotiation—he harbored a deep if not altogether ill-founded suspicion of journalists—he agreed to let me walk with him. He told me that he would be hiking east on TX-73 somewhere outside Winnie, Texas, on a certain day in early June. If I could find him, I was welcome to tag along, but he wasn't slowing down for anybody.

+

On the appointed day, my sister and I drove southeast from Houston, eyes peeled for a walker by the side of the road. As we passed a place on the map called Alligator Hole Marsh, we spotted him: a white apparition on the far side of the highway, walking upstream against the traffic. We circled around and parked on the shoulder some fifty yards up the road. He waved as he drew near. He carried a blue backpack no larger than a preschooler's knapsack. A single plastic water bottle was tied to his belt with a piece of frayed blue string. His trekking poles were folded in the crook of his arm. In his hand, he carried a chipped Styrofoam coffee cup.

When he reached the car, I shook his hand, and he smiled. He had a wild head of white hair streaked with yellow, and a white beard threaded with black. Both reached down to his collar, where they whorled, oceanic. Atop his head he wore a white runner's cap. He took his sunglasses off, and his eyes, arced against the sun, were fixed with deep, leathered creases, pale in their depths. His hands too were deeply tanned, but only up to around the base of his thumb; the rest, shaded by the cuffs of his shirt, was pink.

Any repeated action will create patterns, and after forty-six days on the road, he was embroidered with them. He showed off his beat-up black running shoes, with holes where his toes poked up, the soles slightly sloped from his tendency to pronate. His white button-down dress shirt, which he picked up for fifty cents at a secondhand shop, bore a dark stain from where his pack rode, like a burn shadow.

His real name was M. J. Eberhart. He said I could call him "Eb."

Bright slabs of metal and glass roared past, summoning a hot wind. Eberhart sat down on the rear bumper of my sister's station wagon. We had brought him an ice-cream cone and some cold water, which he accepted bashfully. "Oh, this is such a blessing," he said. "Oh man." A smile lingered on his lips as he slurped at the vanishing cone.

"I've always heard that it's better to give than to receive, but someone has to receive, and I've learned to do it," Eberhart wrote in his hiking memoir, *Ten Million Steps.* The book is filled with stories of people paying for his meals, taking him into their homes, and pressing wads of cash into his hands. He usually protested, and, ultimately, always relented. He received his first trekking poles, a pair of expensive titanium walking sticks from Germany, from a fellow hiker he had known for less than three hours. Both in person and in writing, Eberhart was unfailingly grateful. The words *thank* and *thanks* appear more than one hundred times in *Ten Million Steps.*

(By way of comparison, the book contains only sixty-seven uses of the word *trail*.)

Finishing his ice cream and handing the wrapper back to my sister, Eberhart topped off his water bottle and filled the Styrofoam cup with ice. With that, he was ready to go. I hugged my sister good-bye. She got back in the car and drove off, leaving me and Eberhart alone with a million acres of green ranchland.

"Welcome to my backyard," Eberhart said, waving at the vastness with his cup of ice. The land was flat (elevation: eleven feet), but the clouds above it were colossal—a white mountain range, severed and levitated.

As we walked, Eberhart recounted his travels thus far. He had begun forty-six days earlier at the southern terminus of the Continental Divide Trail. From there, he headed east, through the blackened badlands of New Mexico, through the gateway city of El Paso, and onto an endless spread of dry dun plains haunted by dust devils. The traffic consisted almost entirely of semitrailer trucks, silvery leviathans that surged past every ten seconds at speeds of a hundred miles per hour. He learned to take shallow breaths through his nose, so as to not inhale their fumes. The sound was meteoric.

In West Texas, the highway stretched in a straight line to a vanishing point on the horizon. Space and time started to play tricks on him. He walked for hours each day and never seemed to progress, the distant mountains retreating faster than he could catch them. The highway was lined with mileage markers, and he checked each one to convince himself that the numbers were changing.

Always, there was the wind, which pushed against his forehead during the day and blew sand into his tent at night. In an effort to escape it, one night he set up camp inside an abandoned house in a ghost town, and punctured his inflatable sleeping pad on a shard of broken glass. Another night, he spent slumped over in a booth at a truck stop. The desert mornings were frigid. He began each day hunched over, the hood of his plastic rain poncho pulled up, hands in his pockets.

His plan was to walk from gas station to gas station, but buildings of any kind were sometimes dozens of miles apart. If people hadn't stopped to give him water, he may well have died. When he emerged from the desert, vultures were circling ominously over his head.

Other than the vultures, almost all the wildlife he had seen was dead (most of it roadkill), including a crushed coral snake, two mule-deer, a raccoon, an armadillo, numerous birds, and a group of dead coyotes wired, inexplicably, to a fence.

This experience was not unique to West Texas. Since highways optimize to speed and steel, they have a tendency to kill the slow and soft. As we talked, in short succession, we passed: a turtle, a snake, an armadillo, a baby alligator, and an unidentifiable creature, perhaps a dog, whose fur and bones were fanned out in a radial pattern, as if it had fallen asleep in the shade of a rocket's thrusters.

"We got all kind of roadkill here today," Eberhart remarked.

I asked him if he didn't find the highways an unwelcoming space for a hiker.

"I enjoy being out on the roads," he replied. "You get to see the towns, you get to mingle with the locals. It's a different experience entirely than the green tunnel."

Roads have historically attracted a strange breed of walker—what might be called *gregarious ascetics*. Around the turn of the twentieth century, the poet-vagrant Vachel Lindsay walked thousands of miles across America's roads, having sworn a vow of poverty, celibacy, and sobriety, preaching what he called the "gospel of beauty." Nearly half a century after he published *A Handy Guide For Beggars, Especially Those of the Poetic Fraternity*, a woman named Mildred Norman decided to follow in his footsteps; she changed her name to Peace Pilgrim and began walking coast to coast across the country's roadways promoting a philosophy of nonviolence.* Neither carried any more possessions than would fit in his or her pockets. Both were dependent on strangers for all their food and lodging.

After seven or eight miles, we ducked into a convenience store. The air inside was delicious. The whole room reverberated with a sound at once alien and deeply familiar, a chorus of humming compressors and shushing liquid, percussed sporadically with the clatter of ice cubes and cold coins.

An ancient woman with the face of a baby bird sat behind the counter, propped up in a wheelchair, her bony arms wrapped in blue veins. She greeted us in a hoarse whisper.

"Hey, how in the heck are you?" Eberhart joyfully called out to her. "Do you have a fountain? You *do* have a fountain! Doodah!"

He carried his chipped, dirtied Styrofoam cup over to the soda fountain, where he filled it to the brim with ice and then splashed in some clear liquid. He took a long sip then filled it again. Turning with a contrite expression, he shuffled over to the counter.

"I thought that was water, carbonated. Well, what I done was, I drank about half of it, and then I hit the Sprite button, so if I pay, that's what I'm gonna be stuck with."

"That's okay," the woman rasped. "I ain't charging you."

"It's just getting awful hot out there," he said, apologetically.

"I bet it *is*," she said.

He thanked her sincerely, and we hiked on.

---

* Incidentally, Norman was also the first woman to hike the entire length of the Appalachian Trail (albeit in discontinuous fashion). On her hike, she subsisted off of uncooked oatmeal, brown sugar, dried milk, and whatever wild foods she could gather along the trail. She described her thru-hike as a "toughening process" that prepared her for her lifelong pilgrimage.

"See, that ain't fair," he later confessed to me. "You shouldn't be able to pull people's emotional cords like that. I take advantage of it so much."

+

Bit by bit as we walked I learned the full story of how M. J. Eberhart became Nimblewill Nomad. He was born Meredith Eberhart—which, he stressed, back then was "a *boy's* name"—in a "sleepy" town in the Ozarks with a population of 336. He likened his childhood to that of Huck Finn: He spent his summers running barefoot, fishing, and riding horses. In the fall, he hunted quail with his father, a country doctor.

Eberhart later attended optometry school, got married, and helped raise two boys of his own. They lived in the town of Titusville, Florida ("Space City, USA"), where he was soon making a six-figure salary performing pre- and post-operative work on cataract patients, many of them NASA scientists. He enjoyed helping people restore their sight and he prided himself on being able to provide for his family, but his work still felt oddly hollow. (He was especially irritated by the endless amount of administrative and legal paperwork, which seemed to grow worse every year.)

He retired in 1993 and began spending more time living alone on a plot of land he was developing beside Nimblewill Creek in Georgia. He and his wife started to drift apart. There followed a dark period of about five years, about which he said he didn't remember much. When I later called up his sons—neither of whom had spoken to him in years—they recalled him as a caring father and a dutiful provider, but also someone who was easily frustrated, prone to bouts of drunken brooding, and, occasionally, loud (but never violent) outbursts of rage.

His new house sat near the base of Springer Mountain, which he would regularly climb. His hikes gradually grew longer; he began systematically hiking the AT section by section, eventually reaching as far as Pennsylvania. Then in 1998, at age sixty, he decided to set out on his first "odyssey," a 4,400-mile walk from Florida to Cap Gaspé in Quebec, along a sketchy agglomeration of trails, roads, and a few pathless wilderness areas. Not long before, he had been diagnosed with a heart block, but he declined the doctor's admonitions to have a pacemaker installed. His sons assumed he would not make it home alive.

On the trail, Eberhart renamed himself after his adopted home, Nimblewill Creek. He began in the swamps of Florida and hiked north on flooded trails, where the dark, reptilian waters sometimes reached to his waist. When he emerged from the swamps, all ten of his toenails fell off. By the time he reached Quebec, it was already late October. Over the previous ten months, he had experienced a slow religious awakening, but his faith was shaken as he passed through those grim, freezing mountains. "Dear Lord, why have you forsaken me?" he asked, upon seeing the weather darken one day at the base of Mont Jacques-Cartier. However, a lucky break in the storm allowed him to reach the snowy mountaintop, where he sat in the sun, feeling "the warm presence of a forgiving God." After reaching the trail's end, he returned to the South (on the back of a friend's motorcycle) and, in a blissful denouement, walked another 178 miles from a town near Miami down to the Florida Keys, where he settled into "a mood of total and absolute, perfect contentment, most-near nirvana."

He returned home a different man. He stopped showering. He kept his hair long. He began ruthlessly shedding his possessions; over the course of three days, he burned most of the books he had collected over his lifetime, one by one, in a barrel in his front yard. In 2003, he and his wife divorced. He ceded the house and most of his assets to her, and signed over his other real estate holdings, including the land at Nimblewill Creek, to his two sons in an irrevocable trust. Since then, he has lived solely off his social security checks. If those funds ran out by the end of the month, he went hungry. But what he had gained was the freedom to walk full-time, which felt to him like freedom itself. "As if with each step," he wrote, "these burdens were slowly but surely being drained from my body, down to the treadway beneath my feet and onto the path behind me."

+

Eberhart paused by the roadside to pull out a map, which he kept sealed in a plastic sleeve. The map covered no more than fifteen miles; it took one hundred seventy of these maps to cover the whole trip. He so dreaded the prospect of getting lost that he had begun carrying a little yellow GPS unit as backup, despite the added weight and cost.

I was surprised to find that in addition to the GPS, he also carried a small cell phone, a digital camera, and an iPod touch (which he used to log on to free wireless networks so he could check the weather, publish online journal entries, and answer the occasional email).

The cell phone he carried at the insistence of his girlfriend, Dwinda, an old high school sweetheart, who called twice during our trip to check in on him. Initially, he said, he had regarded it as unnecessary weight, until he broke his leg while hiking in the Ozarks and it ended up saving his life. "I don't gripe about the phone anymore," he said. "I enjoy having it and talking to her every day." He told me that though nobody can "totally escape" the world of technology, you can still "keep it from totally overwhelming you and controlling your life."

"I think I've been pretty successful in that regard," he said.

Around our tenth mile together, Eberhart noticed that the shoulder of the road had begun to narrow. He predicted that we were nearing the city of Port Arthur. Sure enough, the city soon appeared: a jagged row of mechanical spires spouting white steam, which turned and dissipated hypnotically in the blue air.

Port Arthur was a petro-town, known for refining the oil pulled from deep beneath the gulf. ("Where oil and water mix, beautifully," read the Chamber of Commerce's motto.) We passed a series of barbed wire fences, behind which long white tubes shone like bones. The looming oil refineries recalled the Futurist architectural designs of Antonio Sant'Elia. This was where much of the nation's gasoline came from, as well as its plastics and its petrochemicals. I had never seen anything like it before. It was as if we had wandered below the deck of a gleaming cruise ship and found ourselves in the engine room, amid the sooty machinery that kept the ship gliding along.

Cars were lined up, idling, in their rush to get home. The air was sickly with their exhaust. "How do they live like this?" Eberhart said, looking at the pinched faces behind the windshields. "They spend more time sitting around in their vehicles than they do at home!"

At a wide intersection in front of the Valero refinery, we stopped at the window of a squad car to ask for directions. A police officer with a buzzed scalp and two missing teeth sat at the wheel, with a woman in plain clothes sitting beside him.

"Hang a left in about four blocks," he said. "Stay on the main road, though, because you're heading into the worst part of town."

We walked along the sidewalk—a luxury, after fifteen miles of highway shoulder—through a neighborhood of small, tidy houses. In the parking lot of the convenience store, a man sat on the tailgate of his truck drinking a tallboy of beer. Inside the store, it was again clean and cold. To Eberhart's delight, this store had six varieties of frozen burrito. He had been living off of frozen burritos for weeks and had acquired a taste for them. ("If you didn't eat them out west there, you didn't eat, because that's all they've got. Breakfast burritos, lunch burritos, I think they even had a dessert burrito.") We sat near the back of the store, on crates of soda, eating our dinners in the cool air.

As Eberhart was moving on to his dessert course—a half-pint of Blue Bell vanilla ice cream—one of the store's clerks asked him to move to the side so he could restock the refrigerator. Eberhart apologized.

"Everyone else who comes through here has a car to sit in, but we're walking," Eberhart explained.

The store clerk looked suspicious.

"Where you going?" the clerk asked, in the rapid, trilled patter of a native Hindi speaker. Eberhart had a thick, slow Missouri accent that bent "wash" into "warsh." A certain amount of miscommunication ensued; I found myself playing translator between the two.

"Well," Eberhart said, "we're going to go across the bridge tomorrow and into Louisiana. I'm heading for Florida at the end of this month."

"Are you going for a record?"

"No, just walking."

"Just for fun?"

"Well, yeah . . ."

"Where are you living, nighttime?"

"I have a tent for camping."

"You take a bath anywhere?"

"Not as often as you do . . ."

"How many days?"

"I started in New Mexico forty-six days ago."

The man paused and cocked his head. "What is your reason?"

"Well, I'm a long-distance hiker, and I enjoy walking. Meet people. Have some ice cream," Eberhart chuckled.

"Yes, yes."

"It's not a bad life. Sometimes when it storms real hard you get wet . . ."

Eberhart fished around in his wallet and pulled out a business card, on which a red line marked the entire route around the continent. The store owner still looked perplexed. "You should tell the media," he said. "You could be in the local paper."

Eberhart's smile tightened.

Later, Eberhart told me that these questions were exceedingly common. He understood why people asked them; they saw him as "a total alien." Naturally, they were curious. However, the one question he dreaded was the simplest: *Why?* "You can answer questions all day, but you just don't want to answer *that* question," he said. "You know why? Because you *can't* answer the damn question."

+

Why *do* we hike? I have asked many hikers this same question, and I have never received a fully satisfying answer. It seems there are many overlapping reasons: to strengthen our bodies, to bond with friends, to submerge ourselves in the wild, to feel more alive, to conquer, to suffer, to repent, to reflect, to rejoice. More than anything, though, I believe what we hikers are seeking is simplicity—an escape from civilization's garden of forking paths.

One of the chief pleasures of the trail is that it is a rigidly bounded experience. Every morning, the hiker's options are reduced to two: walk or quit. Once that decision is made, all the others (when to eat, where to sleep) begin to fall into place. For children of the Land of Opportunity—beset on all sides by what the psychologist Barry Schwartz has called "the paradox of choice"—the newfound freedom *from choices* comes as an enormous relief.

This form of freedom is a curious thing, at once an expansion and a constriction of one's options. Ever since America declared its independence from England, it has framed itself as a home for lone, free actors. Walking has always symbolized and enacted this untethered state. Thoreau memorably wrote that only if "you are ready to leave father and mother, and brother and sister, and wife and child and friends, and never see them again; if you have paid your debts, and made your will, and settled all your affairs, and are a free man; then, you are ready for a walk." It's this reduction that makes a walk feel so freeing. In walking, we acquire *more* of *less*.

+

We slept that night on a grassy berm a few blocks from the convenience store. I strung up my hammock between an electrical pole and a chain-link fence bearing a sign that read: WARNING: LIGHT HYDROCARBON PIPELINE. As the sun set, orange and green balls of light appeared, floating, amid the dark refineries. Above one tower in the distance rose a pillar of flame.

We awoke the next morning to sirens and songbirds. A heavy dew coated the grass, my hammock, his tent. The air was aflame with mosquitoes. "All right new day, here we come," Eberhart said, stretching out his arms. "This old buzzard's upright one more day."

Back on the road, he started off slowly. He said it took him at least a half hour to get the kinks out of his back. "Every morning, you think, oh man, this isn't going to go away today," he said. "But it does."

As we passed the refinery again, he stopped to take a picture. "That feller over there makes its own clouds, doesn't it?" he said. Among the hikers I had met, Eberhart showed a rare appreciation for human environments. When taking landscape photos, most hikers will go to great lengths to crop out any power lines. But Eberhart said that once you resign yourself to the fact that every photo is going to have power lines in it, you'll find you can take much better pictures.

The problem, he said, was that hikers tended to divide their lives into compartments: wilderness over here, civilization over there. "The walls that exist between each of these compartments are not there naturally," he said. "We create them. The guy that has to stand there and look at Mount Olympus to find peace and quiet and solitude and meaning—life has escaped him totally! Because it's down there in Seattle, too, on a damn downtown street. I've tried to break those walls down and de-compartmentalize my life so that I can find just as much peace and joy in that damned homebound rush-hour traffic that we were walking through yesterday."

As a proud traditionalist, Eberhart was startled when I told him that, in this regard, he was in the vanguard of environmental philosophy. In the last twenty years, postmodern environmental scholars like William Cronon have subverted the gospels of Muir and Thoreau by arguing that nature, as a world distinct from the realm of human culture, is an empty and ultimately unhelpful human construct. The concept of nature cleaves the planet in two: presuming that there is a natural world over here and a human world over there. Cronon argues that this division not only alienates us from our own planet, it also obscures our origins as animals, as collections of cells, as collaborative and intertwined living beings. All organisms are involved in a constant process of reshaping our world to better suit their needs, whether they are termites building mounds, elephants felling trees, kudzu vines colonizing an abandoned house, or shepherds and sheep working together to trample out a grassy turf. The thing we call nature is in large part the result of these minute changes and adaptations. There is no single entity or primal state one can point to and call "nature": it is both everything and nothing.

There are many smart and conscientious people who argue that humans should resume living in a more natural way, which we have abandoned at some point in the past. But the problem with equating the natural with the good, Cronon argues, is not simply that the concept of nature is illusory—it is also counterproductive. "When we speak of 'the natural way of doing things,' we implicitly suggest that there can be no other way," he observed. By arguing that something is natural—and thus "innate, essential, external, nonnegotiable"— we short-circuit any meaningful conversation about how the world should be.

At the eastern edge of Port Arthur, we crossed a bridge, which arched over the river to reach a green island. On the far side of the bridge stood a sign reading WELCOME TO PLEASURE ISLAND. There were sailboats, fishing piers, a golf course, a castle made of wood, and an RV park. We walked until we reached a bait shop, where we stopped for breakfast. Eberhart handed the perplexed cashier his business card. With her permission, we spread our wet rainflies out on the front yard. As they dried, we sat together at a picnic table in the shade. Eberhart ate a cheese danish and drank a cup of coffee. He looked deeply content.

"You know what, Rob?" he said, unprompted. "You can be a whole lot happier if it don't take a whole lot to make you happy."

He thought for a moment.

"You know when you see a child, and you see that innocence, that glow in a child's face? It goes away when we grow out of childhood, and it doesn't come back almost until you're old and feeble. But you still occasionally see it in someone. They don't even have to speak. You see it in their face; they have that inner joy, that inner peace. I hope you can see some glimmer of that in my face."

His sunglasses were off, and he was staring me in the eye, earnestly. "So is it real or isn't it, when you look at me?" he asked.

I looked at him closely, mildly unsettled by the directness of his question. There was indeed something radiant in him—his cheeks were glowingly pink, his eyes guileless and clear. But I also thought I detected something else, the faint smolder of an ancient anger, carefully suppressed. What peace there was seemed so fragile, so incipient, that I was wary of prodding it.

I dodged his question with another question: Was peace a lasting state of being, I asked, or was it something you had to continually work on?

"You work on it day to day," he said. "It's a daily pilgrimage, something that you renew. I would be miserable today if I let myself. And the peace that I've enjoyed these last number of years is fleeting. It could be gone in an instant."

We packed up our rainflies, which had grown crisp in the sun. When we left, our packs felt lighter. The road entered a swampy

marsh. Wind invisibly petted the wet green grass. Eberhart stopped
and bowed his head to recite a prayer he had written. He adopted
the cadence of an old cowboy poet, his voice misty with sincerity.

Lord, set me a path by the side of the road,
Pray this be a part of your plan.
Then heap on the burden and pile on the load,
and I'll trek it the best that I can.
Please bless me with patience; touch strength to my back;
Then cut me loose and I'll go.
Just like the burro toting his pack,
The oxen plowing his row.
And once on this journey, a witness for you,
Toward thy way, the truth, and the light.
Shine forth my countenance steady and true,
O'er the pathway to goodness and right.
And lest I should falter and lest I should fail,
Let all who know that I tried.
For I am a bungler, feeble and frail,
When you, dear Lord, I've denied.
So blessed be the day your judgment comes due,
And blessed be thy mercies bestowed.
And blessed be this journey, all praises to you,
O'er this path by the side of the road.

The prayer, he said, had come to him one day on his transcontinental hike from Cape Hatteras, North Carolina, to Point Loma, California. He spoke it into a microcassette recorder that he used to carry,
and transcribed it when he got home. He was fond of saying that he
paused only to look up the words he didn't know in the dictionary.

As we approached the Louisiana state line, the air began to prickle
with heat. Eberhart's shirt had turned a diaphanous shade of pink. At

one of the convenience stores we passed, I filled up my water bottle with tap water. Outside, when I opened it to take a sip, I recoiled: it smelled like kerosene. Eberhart said that was common around here. "The last two or three places, they said, 'Oh man, you don't want to drink this stuff,'" he said. "In fact that one lady back there said, 'Just go over there to the cooler and get yourself a bottle of water.' Some of this stuff tastes pretty bad."

Curiously, for a man who spent nearly all his waking hours outdoors, this pollution did not seem to greatly anger Eberhart. He wasn't in support of pollution, of course, but he was skeptical of efforts to tighten environmental regulations any further. To him, personal freedom was sacrosanct, and anything that impinged on that freedom was dangerous.

"God put us here on this earth so that it can be productive and provide sustenance," he explained. "Petroleum is a natural resource. So yeah, we're going to use it up, eventually. Before that day comes, hopefully, they'll bust the hydrogen molecule, and then, damn it, we're going to the moon again! But until that day, what are we going to do? Go back and get the plow and the horse and build a cabin and burn firewood and fight off the Indians? What are we going to do? The Lord *put* these resources here for us."

I was surprised, and I told him so. Based on his minimalist lifestyle, I'd assumed he would recommend a lighter footprint for other people as well.

"I'd be happy to talk to you about it, but I'm not going to force my lifestyle on you," he replied. "That's what's happening. These things are being forced on me. I resent it. I absolutely resent it. If I want to buy an airplane and fill it full of a thousand gallons of fifty-dollar-a-gallon fuel, and I got the money to do it, goddamn it, leave me alone!"

A lengthy—at times, heated—discussion about the proper role of government regulation ensued. As we continued talking over this

point, though, I discovered our disagreement was not so much political as epistemological; far from agreeing on the solution to pollution,
we could not even agree on the existence of a problem. Environmental science, Eberhart believed, was "politicized and bastardized"
by people with a dual agenda: to increase the government's control
over its citizens and the desire to put man above God. Darwinism,
he argued, was also part of this ruse. He waved his hand at the land
around us. "If you stand here and look out here across these fields,
you can see that there is an order. You say that's all just chaos, but it
isn't. There is an order. And that didn't come from Darwin."

I looked around. On either side of the road was tall marsh
grass, which spilled lushly through a barbed wire fence, pushing
the posts askew. That grass housed countless insects and reptiles
and birds. When Eberhart looked out over those fields, he saw a divine creation, infused with purpose, perfectly designed and lovingly
tended. Whereas when I looked at the same field, I saw the miracle
of evolution—an infinitely complex constellation of particles, cells,
bodies, and systems competing and cooperating, reproducing and
dying off, self-perpetuating and yet always in flux. I tried hard to
imagine the field from his perspective, and then again from mine.
Both were beautiful—even awe-inspiring—in their own ways.

"I believe in the Maker, I believe in the hereafter, I believe in the
Holy Spirit. I believe in these things as deep in my heart as anyone
could," he said. "Now, in all this discourse, I haven't told you that
you're wrong. You can believe what you want, but I have the right
to believe what I want. What do you expect from an old man who
was raised in a church-based community, where right was right and
wrong was wrong?"

By this point the afternoon heat had closed in around us like a
bad hug. Sweat flies tangled, vibrating, in Eberhart's beard. It grew
too hot to argue, and we still had miles more to walk, so we fell into
an uneasy silence. We were both relieved to reach the next stop, a

truck-stop restaurant, where we sank into cool plastic chairs. A family of four gave us their leftover onion rings, fries, and a half-eaten burger. Five miles later, we stopped at another oasis of air conditioning. At the table next to us sat a mother with her two rambunctious young sons. The boys silently gaped at Eberhart as he walked past, then promptly forgot all about him and returned to torturing each other.

Their exhausted mother, who helped run the store, told us that this area had been hit by two hurricanes since 2005, and the store had been flooded both times. They weren't able to get flood insurance anymore, so each time a hurricane approached, they simply packed all the merchandise into a U-Haul and tried to outrun it.

"But you keep coming back?" Eberhart said, wonderingly. "That is incredible."

She explained that the hurricanes were unpredictable: "They hit and miss." The last hurricane had left her bicycle right where she had left it on the lawn, but had hurled an old washing machine into the marsh behind her house. Sitting there, I pondered how it must feel to live in the shadow of such roiling, ravening complexity. I wondered if it would make me believe in a vengeful and capricious God, or no God at all.

I once read a study that found that in this very parish of Louisiana—which is predicted by scientists to be severely damaged by rising sea levels and increased storms—more than half the pop-ulation disagreed with the scientific consensus that human activity was altering the climate. Eberhart too was skeptical. ("It's one hell of a leap of faith," he'd said.) Indeed, when I considered it here, where the hurricanes raged worst, I was momentarily amazed by the idea that we tiny humans are capable of radically altering the planet, when for our whole existence it has tossed us around like so many ants. It was as if we had wounded one of the gods.

We are all children of landscapes. We first learn about the world in the place where we grow up; it shapes our language, our beliefs, and our expectations. I was raised on the shores of Lake Michigan, among plunging ravines and manicured lawns, where human ingenuity reversed the flow of the Chicago River and converted the prairies into corn and concrete. Eberhart was from the northeastern Ozarks, a hard land, trod by Meriwether Lewis and terrorized by Jesse James, where lead ore was once pulled from the ground and hauled off by oxen until it ran out, leaving the people to scrape out a living from the acidic soil. This woman and her children were from a place where, any summer or fall, with little warning, a cloud could erase their lives.

The sun was setting by the time we left the store. We walked down the roadside searching for a dry place to sleep. First we looked behind a firehouse on stilts, then behind a cemetery named Head of the Hollow. On the far side of the graves, Eberhart spotted a motte of live oaks. He hopped nimbly over the barbed wire fence and ducked beneath their gnarled branches. "Beautiful!" he called out.

Inside was a shady grove, carpeted with slender, rippling leaves. We hurriedly set up camp and crawled into our nylon cocoons as mosquitoes descended like mist. I thought back to what Lamar Marshall had said, that Eberhart's way of walking "defeats the whole damn purpose of being in the woods." Here we had discovered a corner of wilderness that few people ever slept in, and it was truly lovely. "A person with a clear heart and open mind can experience the wilderness anywhere on earth," the poet Gary Snyder once said. "The planet is a wild place and always will be."

+

Wilderness, according to William Cronon, is as illusory a concept as nature. He writes that wilderness is too often seen as an Edenic escape from the modern world—"the tabula rasa that supposedly existed before we began to leave our marks on the world." It is both our fantasy and our fallback plan; we forgive the poisoning of our local waterways so long as Yellowstone remains pristine. "By imagining that our true home is in the wilderness," he writes, "we forgive ourselves the homes we actually inhabit."

But Cronon argues that unlike the concept of nature, which tends to narrow our thinking, wilderness can broaden it. In the wild, we witness firsthand that there is a world of stunning complexity that existed prior to us and will always stubbornly resist our attempts to simplify it. "In reminding us of the world we did not make," Cronon writes, "wilderness can teach profound feelings of humility and respect as we confront our fellow beings and the earth itself." (This lesson, Cronon writes, "applies as much to people as it does to (other) natural things." So his "wilderness" and Scheler's "fellow feeling" are, oddly enough, kindred concepts.)

On a farm, the land is narrowly defined by how it profits the farmer—he sees little more than crops, soil, storm clouds, pests, debt. But the defining feature of wilderness is its unruly condition: it is the land we leave to *grow wild*. The wilderness has always been defined as the land out there, beyond the fence, *not-self, not-home*. It is open land, which no one owns, and no one can claim to fully know. Throughout history it has offered a home to all manner of prophets, explorers, ascetics, outcasts, rebels, fugitives, and freaks. Some, like Muir, found it holy; others, like the Puritans, found it horrible. None, however, could hope to fully grasp it; it is forever *beyond us*. Perhaps this is why the wilderness, as sung into being by Thoreau—"this vast, savage, howling Mother"—has managed to retain its transcendent power in our increasingly secularized and post-natural society. Whether it be a snowy mountaintop or a shady grove, the wild is a place where both Eberhart and I, different as we may be, could feel bathed in the same cosmic light.

On wild land, wild thoughts can flourish. There, we can feel all the ragged edges of what we do not know, and we make room for other living things to live differently. Cronon boldly concludes his essay on wilderness by asserting that we must learn to reinfuse this sense of the wild back into the human landscape—for instance, to see even the trees in our backyards as wild things—and to reframe our understanding of the wilderness so that it can contain us within it. The next great leap in our ecological consciousness, he argues, would be to "discover a common middle ground in which all of these things, from the city to the wilderness, can somehow be encompassed in the word 'home.'"

+

The following day was my last with Eberhart. A little after dawn we snuck back to the road and resumed walking eastward. Hot gray skies pressed earthward. Somewhere, a marsh was burning. To our left were vast ranchlands. A helicopter hovered low over them, releasing a fine chemical mist. Off to the right, bulbous storm clouds floated in from the gulf, trailing gray tentacles.

In ten miles, at the storm-ravaged town of Holly Beach, I would hitch a ride back to Houston with a group of touring Danes, rewinding, in an hour, all the progress we had made in three days. But that morning, the end still seemed a long way off, and my legs had already begun to ache. Walking on the uniform surface of a road for days wears on the body in the same way working in a factory does. The same motion is repeated, with very little variance, thousands of times each day. Odd body parts grow sore: the backs of the knees, the bottoms of the feet.

Eberhart, meanwhile, seemed fine. His posture was hunched, and he had a slight hitch in his right step, but his stride, from the outset, was remarkably steady: three miles an hour, on the tick. From time to time he stopped, leaned forward on his trekking poles, and swung first one leg backward, then the other, to loosen them up. Throughout the day, to ease his pains, he swallowed handfuls of aspirin and joint supplements.

At his age, after all he had experienced, it was amazing he could hike at all. On his journeys, he had broken four ribs, his shinbone, and his ankle. He had suffered from excruciating bouts with shingles and an abscessed tooth. He had visited unspeakable horrors upon his feet. Once, up in Canada, he had been struck by lightning. To help me understand the sensation, he asked me to imagine being soaked in gasoline and then touched by a lit match. "It goes VOOSH," he said. "There's no vibration or nothing. It just passes through you."

He told me that when he was a young man, he had been taller than me, standing almost six feet tall, but over the years his spine had compressed. "I'm shrinking," he said. "My body is shrinking, my mind is shrinking, my vocabulary is shrinking. My ability to maintain a thought sequence, it's not *gone*, but it's not like it was ten years ago. Part of the thing is . . ." he paused. "Look at that flatware!" he exclaimed as he bent down to pick up a crippled fork. "Look at that! Is that beautiful?"

One of his hobbies, he told me, was to collect discarded silverware along the side of the road. He said he hoped to one day put together a full "flatware set" of flattened utensils, eight of each.

All along our hike, he had picked up other shiny objects as well: coins, keys, marbles, car wash medallions, hearing aid batteries. When he reached the next post office, he would mail all that he had collected back to his sister's house, where he stored his findings in two Mason jars.

This habit of picking up jetsam on the side of the road fascinated me. It was perhaps the starkest of his many ironies. In virtually all other respects, he was a fanatical minimalist. Even at home, he saw these, the last years of his life, as a process of winnowing. He owned scarcely more than he could fit in his truck. In his sister's basement was also a cardboard box full of mementos, photographs, a few sentimental objects that had belonged to his parents. He said he was struggling to work up the nerve to get rid of those as well, but shedding childhood attachments was "a tough, tough process."

"I tell my friends: Every year I've got less and less, and every year

I'm a happier man. I just wonder what it's going to be like when I don't have anything. That's the way we come, and that's the way we go. I'm just preparing for that a little in advance, I guess."

A few minutes later, Eberhart paused at the intersection of a gravel road to show me the contents of his pack. He spread out his things in the dust. There was a tarp tent, a sleeping bag, a sleeping pad, the small bag of electronics, a hint of a medical kit, a plastic poncho, his maps, a pair of ultralight wind pants, and the pile of metal junk. All the fabrics had the wispiness of gossamer; a strong wind could have taken most of his earthly possessions away.

To cook his meals, he used to rely on a tiny wood-fired stove of his own invention, but he had since ditched that. He listed some of the other things he had brought on his first thru-hike but later discarded. He had traded his heavy leather boots for trail running shoes. He exchanged a three-pound internal frame pack for an eight-ounce frameless one, and a three-pound synthetic sleeping bag for a one-pound down bag (with the zippers trimmed off). Instead of a toothbrush, he carried a wooden toothpick. He did not carry a spare change of socks, a spare set of shoes, or any spare clothes. He did not carry any reading material, or even a notebook. He did not carry toilet paper. (Instead, he used the subcontinental rinse-and-rub method. When water was scarce, he rinsed with his own urine, which he then cleaned off with a careful splash of water.) His medkit contained little more than a few Band-Aids, a pile of aspirin, and a sliver of a surgical blade.

Shaving down one's pack weight, he said, was a process of sloughing off one's fears. Each object a person carries represents a particular fear: of injury, of discomfort, of boredom, of attack. The "last vestige" of fear that even the most minimalist hikers have trouble shedding, he said, was starvation. As a result, most people ended up carrying "way the hell too much food." He did not even carry so much as an emergency candy bar.

Earlier, I had asked him if he was afraid to die. He shook his head. "Nah, I don't think so," he said. He told me his grandfather had died in the woods (of a heart attack while hunting), his father died in the woods (of a chainsaw accident while gathering firewood), and he was "working on it."

"I threw my fears and worries away a long, long time ago about being out in the wild," he said. "I've been out there so long and so far, by myself, and never felt more at peace and more secure and more in my element. It's not an adrenaline pump or anything like that. It's a resignation just to let it be the way it's going to be."

As I picked over his gear, one question kept nagging at me. Feeling sheepish, I asked if the rumor I'd heard was true: Did he have all his toenails surgically removed?

He smiled. "Oh, sure," he said.

He sat down and pulled off his tattered sneakers, and then peeled off his socks. His ankles were a shocked shade of pale below the sock line. His pink toes, rimmed with yellow calluses, were long and knobby. When I looked closer, I saw that it was true: They had no nails, except for a few whiskery fibers that were trying to grow back.

He said that whenever people questioned his dedication to the life he had chosen, or tried to downplay his journeys as a mere lark, he would pull off his shoes and show them his feet.

"Can you imagine what it's like to have all your damn toenails ripped out at the roots and then have acid poured on them so they won't grow back?" he said. "Do you have any idea what that feels like? You think that's a lark?"

+

Back home, I tracked Eberhart's progress from the journal entries he periodically posted on his website. He reached the end of his walk one night in Florida, where he knelt beneath a streetlight and said a prayer of thanks. He had told me that this would be his last thru-hike. But the following summer, he was back on the road, walking the full length of the Oregon Trail, followed, in subsequent years, by the California Trail and the Mormon Pioneer Trail. The last time we spoke, he said he was planning to finish hiking the Pony Express Trail, from Missouri to California. On and on he'll go, as long as his feet will carry him.

As one of Eberhart's favorite poets, Robert Service, once wrote:

> *The trails of the world be countless, and most of the trails be*
> *tried;*
> *You tread on the heels of the many, till you come where the*
> *ways divide;*
> *And one lies safe in the sunlight, and the other is dreary and*
> *wan,*
> *Yet you look aslant at the Lone Trail, and the Lone Trail lures*
> *you on.*
> *And somehow you're sick of the highway, with its noise and its*
> *easy needs,*
> *And you seek the risk of the by-way, and you reck not where it*
> *leads . . .*
> *Often it leads to the dead-pit; always it leads to pain;*
> *By the bones of your brothers ye know it, but oh, to follow*
> *you're fain.*
> *By your bones they will follow behind you, till the ways of the*
> *world are made plain.*
> *Bid good-by to sweetheart, bid good-by to friend;*
> *The Lone Trail, the Lone Trail follow to the end.*

Han-shan too wrote honestly about not only the glories of the simple life, but its hardships as well. He bemoaned his crippled body, salivated over the lavish food he had renounced (roast duck, fried pork cheek, steamed baby pig in garlic), and wept over dead friends. Like Eberhart, he left his wife and son so he could roam freely. The ramifications of that decision reverberate throughout his writing; he fondly recalls the sound as his infant son "gurgles and coos"; he has haunting dreams where he returns to his wife, only to find she no longer recognizes him. A chilly sense of regret creeps into even his sunniest remembrances. "How could I know beneath the pines / I would hug my knees in a frigid wind?" he asks.

Reading these lines, I can't help but think of Eberhart, nearing his eighth decade on this planet, sleeping on the hard ground. ("Oh, my arthritic, bony old body," he wrote in his journal one cold desert night in West Texas. "I'll be listening to it complain, for sure.") This is what is left when the haze of romance has burned away. This is the cost of freedom. Every year, the lone, lean life grows harder. "Up high," Han-shan wrote, "the trail turns steep." And yet he went on climbing.

I had gone in search of Eberhart, the modern nomad, to see what my life might have looked like if I had chosen to pursue the life of simplicity a long trail affords. Walking with him, I witnessed both the advantages and disadvantages of a life honed down to a single point: the finer the edge, the more brittle the blade. Eberhart had opted for the path toward maximal freedom, which meant shunning comfort, companionship, and security: he may have to sleep on the ground, but he can sleep anywhere he likes. If he gets sick or injured, he may well die, but, he figures, at least he'll die outdoors.

"It is pleasant to be free," wrote Aldous Huxley, who, like Eberhart, for years owned little more than an automobile and a few books. "But occasionally, I must confess, I regret the chains with which I have not loaded myself. In these moods I desire a house full of stuff, a plot of land with things growing on it; I feel that I should like to know one small place and its people intimately, that I should like to have known them for years, all my life. But one cannot be two incompatible things at the same time. If one desires freedom, one must sacrifice the advantages of being bound."

Freedom, in other words, has its own constraints. Snipping what

Han-shan called "the world's ties" can come as a relief—from the demands of a job, the constant upkeep of a house, even the obligations of friends or family—but those same ties are often what give our lives meaning and provide a buffer against calamity. Sacrifice is unavoidable.

Whether or not we agree with their choices, the dedication these men have to living a free life presents us with an unsettling question: What do we value above all? Is there anything we hold as dear as Eberhart and Han-shan hold their freedom? And to gain it, what would we be willing to lose? What wouldn't we? And then what does that tell us about what we *really* value above all?

+

Old age brings with it another kind of liberation: freedom from the doubt, angst, and restlessness of youth. The old can look back and see their decisions as a single concatenation, sheared of all the ghostly, untaken routes. Heidegger, a forest-dwelling philosopher enchanted with the earthy wisdom of the *Feldweg* (field path) and the *Holzweg* (wood path), discussed his life in this manner. Three years before his death, he wrote to his friend Hannah Arendt: "Looking back over the whole path, it becomes possible to see that the walk through the field of paths is guided by an invisible hand, and that essentially one adds little to it." But he was able to make that judgment only with the benefit of hindsight. Fate is an optical illusion. From the vantage of a thirty-year-old like me, life's path still bristles with spur trails and possible dead ends.

And so we return, once again, to the essential question: How do we select a path through life? Which turns should we take? To what end?

To be able to answer these questions, deftly and with foresight, is what we mean when we say someone is *wise*. Wisdom—not intelligence, not cleverness, not even moral goodness, but wisdom—is what guides us through the unknown. Perhaps the word *wisdom* sounds hoary to your ear. (Indeed, it does to mine.) As the philosopher Jim Holt has written, it has fallen out of vogue with philosophers in recent decades; the *Routledge Encyclopedia of Philosophy* notes that "wisdom has come to vanish almost entirely from the philosophical map." Ancient philosophers defined wisdom as a way to "maximize the good." However, contemporary philosophers have shied away from discussing wisdom, because they view it as an overly "value-saturated concept." "It is not that philosophers are daunted or bored by wisdom," Holt writes. "Rather, they have concluded that there is no single right balance of elements that constitutes 'the good life for man,' and hence no unitary value that wisdom can help us maximize."

He goes on:

> Suppose you are torn between dedicating your life to art (say, by becoming a concert pianist) or to helping others (say, by going to medical school and joining Doctors Without Borders). How do you decide? There is no common currency in which artistic creation and moral goodness might be compared; these are but two of a plurality of incommensurable values that can be realized in a human life. Do you then ask yourself which choice will bring you greater future happiness? That's no good either, for the path you choose will shape the very person you become, along with the preferences you develop; so to base your decision on the satisfaction of those preferences would be circular.

It is time, I believe, to return to the question of wisdom, but to approach it from a new angle. As Holt points out, wisdom is a notoriously difficult concept to define, but I think we can safely describe it as *a time-tested means of choosing how to live.* The element of time is essential. There is a valid reason that, across millennia and across cultures, wisdom has always been considered the province of old people and old books. Likewise, it is no coincidence that many of the transcultural markers of human wisdom (patience, equanimity, foresight, compassion, impulse control, an ability to reside in uncertainty) are exactly those qualities which children notably lack. Wisdom is a rarified form of intelligence born of experience, the result of carefully testing your beliefs against reality. You make an attempt at solving a problem, and sometimes you stumble upon success; other times you make mistakes, and then you correct them. Over time you learn, you adapt, you grow. In other words, wisdom is a form of judgment that *evolves*.

The notion that wisdom must contain subjective values (must prize, for example, moral purity over artistic virtue, or vice versa) is a specious one. Wisdom is the means by which entities reach their varied ends—by which they gain power or create beauty or help others. Wisdom is structural, not ethical. (Machiavelli, for example, was one very wise, very unethical *figlio di puttana*.) In fact, I would argue that all very old things attain a certain kind of wisdom. There is, if we were to look closely enough, a wisdom of trees and a wisdom of seagrass, a wisdom of mountains and a wisdom of rivers, a wisdom of planets and a wisdom of stars. This book, in its admittedly oblique and winding way, has been a search for the wisdom of trails. It is the wisdom required to reach one's ends while making one's way across an unknown landscape, whether it be a sandy seafloor, a new field of knowledge, or the full expanse of a human life. It is deeply human, this wisdom, deeply animal—and it has tremendous bearing on our personal and collective future.

Wisdom is measured by function. Trails that fulfill the needs of their walkers get used, and used trails persist. Those qualities that lead to greater use, and greater longevity, naturally become the essence of a trail's wisdom. One of the reasons that trails are so relevant to the modern human condition is that they are fervently open-minded: a wise trail can go anywhere and carry anyone. However, every trail is not as wise as every other trail; nondiscrimination is not the same thing as radical relativism. Any forest walker can tell you this—some trails simply *work better*. We can then wonder: Are there some qualities that wise trails—the trails that get better with time—hold in common?

I will venture a guess, which I hope will be improved by others in the future. What unites the wisest trails, I have found, is a balance of three values: durability, efficiency, and flexibility. If a trail has only one of these qualities it will not persist for long: a trail that is too durable will be too fixed, and will fail when conditions change; a trail that is too flexible will be too flimsy, and will erode; and a trail that is too efficient will be too parsimonious, and so will lack resilience. The pheromone trails that ants make, for example, brilliantly balance durability and flexibility: they last long enough to lead other ants to a food source but fade quickly enough to allow new paths to form. And ants invariably find the most efficient route, but then wisely temper that route with redundant detours, ensuring a backup plan in case the best route suddenly fails. The result is a trail that is not just time-tested, as we often say, but *world-tested*. Or perhaps even more precisely, the trail is in a constant state of world-test*ing*, adapting to the world even as its conditions change.

Without ever naming it as such, humans have been putting the wisdom of trails to great use since our inception as a species. Science, technology, storytelling: all masterfully exploit the supple wisdom of trails. Our many forms of understanding the world resemble nothing so much as the trail-wise problem-solving of ants: We test multiple theories against the complexity of the world, and then pursue those that work. The better routes last, the worse ones erode, and little by little those that work improve.

It is in this trail-wise manner that we most effectively navigate a world of forking paths. Holt's hypothetical person, for instance, could conceivably research both paths before making a fateful decision. She could pursue both ideals, making forays into each field (long hours spent at the piano; introductory classes in medicine), to suss out which better suits her abilities and proclivities. Holt astutely points out that her goals and values will shift depending on which path she takes. So let her explore both, and see how each begins to shape her. In case she fails at one pursuit or finds it unsatisfying, she can leave herself open to pursuing the other, or some new offshoot she may have never otherwise discovered. Wisdom often wanders: St. Augustine, Siddhārtha, Li Po, Thomas Merton, Maya Angelou—the insight of each was deepened by a wild and meandering youth. "Seeking and blundering are good," wrote Goethe, "for it is only by seeking and blundering we learn."

Indeed, *some* blundering is good. But a lifetime of blundering— to be condemned to a pathless wilderness—would be a nightmare. Fortunately, we do not wander alone. This is where the other half of a trail-wise way of life comes to the fore: The brilliance of trails stems from the fact that they can preserve the most fruitful of our own wanderings, as well as the wanderings of others; then, as those paths are followed, their wisdom further improves and spreads. Likewise, through collaboration and communication, personal wisdom is transformed into collective wisdom.

As the author and activist Charlotte Perkins Gilman wrote in 1904, the thrust of human history thus far has been to develop "lines of connection," which ultimately add up to what she called the "social organism":

> Watch the lines of connection form and grow, ever thicker and faster as the Society progresses. The trail, the path, the road, the railroad, the telegraph wire, the trolley car; from monthly journeys to remote post-offices to the daily rural delivery; thus Society is held together. Save for the wilful hermit losing himself in the wilderness, every man has his lines of connection with the others; the psychic connection, such as "family ties," "the bonds of affection," and physical connection in the path from his doorstep to the Capital city.

The social organism does not walk about on legs. It spreads and flows over the surface of the earth, its members walking in apparent freedom, yet bound indissolubly together and thrilling in response to social stimulus and impulse.

More than a hundred years later, these words have proven surprisingly prescient. Gradually, our collective intelligence has grown—beyond communities, beyond countries, beyond even our own species. Every day, humans carry on conversations across oceans and intuit the intentions of other organisms, weaving a diverse array of needs into a broader plan of action; in this way, we are slowly transcending ourselves. At the same time, our ability to alter the environment—to change the chemical makeup of the sea and sky, to snuff out whole ecosystems—is growing radically as well. The question remains whether the growth of our collective wisdom can keep pace with our capacity for destruction, whether all of us—"walking in apparent freedom, yet bound indissolubly together"—can cooperate to reach our mutual aims.

Over the course of millennia, our first tentative trails have sprawled into a global network, allowing individuals to reach their ends faster than ever before. But one unintended consequence of this shift has been that many of us now spend much of our lives within a world made up of little more than connectors and nodes, desire lines and objects of desire. The danger of such a blinkered existence is that the more effectively these trails deliver us to our ends, the more they can insulate us from the world's complexity and flux, which results in structures that are dangerously fragile, fixed, or myopic. No matter how vast our collective wisdom grows, we would also be wise not to forget how small it is in comparison to the broader universe. "The attempt to make order out of disorder and chaos, *tohu va vohu*, is the essence of every human life," a wise old man named Baruch Marzel once told the essayist David Samuels. "But stories are never the truth. The truth is chaos."

The old man is right, but this is only half the story. The on-tological truth—the deep reality of the world—is chaos. But the pragmatic truth—the truth we can actually use, the truth that leads us somewhere—is chaos refined. The former is a wilderness, the latter is a path. Both are essential; both are true.

+

Han-shan died more than a thousand years ago, yet we know what little we do about him because, throughout the seven decades of his hermitage, he wrote hundreds, perhaps thousands, of poems. Forgo-ing paper, he scribbled his thoughts on trees and rocks and cliffs and the walls of buildings, sometimes, one imagines, graffitiing descrip-tions of the landscape directly onto the landscape itself. (Some three hundred of these poems were ultimately transcribed and preserved by imperial officials.) In a poem written late in his life, Han-shan recalled visiting a village he'd once lived in seventy years prior. All the people he'd known were now dead and buried. Only he was left. The poem concludes with a proclamation:

> here's a message for those to come
> why not read some old lines

I like to imagine this maxim scrawled onto the wall of an abandoned hut in the village. How baffling it must have seemed to passersby. On the one hand, the words are slightly absurd: writing that people should read more is like using your one magical wish to wish for more wishes. But in the end he's right: What else do we have to guide us through this life but—in Han-shan's perfect phrase—these "old lines"?

As Nietzsche once wrote: "The happiest fate is that of the author who, as an old man, is able to say that all there was in him of life-inspiring, strengthening, exalting, enlightening thoughts and feelings still lives on in his writings, and that he himself now only represents the gray ashes, whilst the fire has been kept alive and spread out. And if we consider that every human action, not only a book, is in some way or other the cause of other actions, decisions, and thoughts; that everything that happens is inseparably connected with everything that is going to happen, we recognize the real *immortality*, that of movement— that which has once moved is enclosed and immortalized in the general union of all existence, like an insect within a piece of amber."

We are born to wander through a chaos field. And yet we do not become hopelessly lost, because each walker who comes before us leaves behind a trace for us to follow. The full span of trail-making on earth, in its broadest sense—all the walks, all the stories, all the experiments, all the networks—can be seen as part of a great communal yearning to find better, longer-lasting, more supple ways of sharing wisdom and preserving it for the future. Ultimately, Hanshan's genius, born from a life spent wandering and pondering trails, was to realize that inherited wisdom can take us far, but only so far. After that, we must explore all on our own.

When I started this book, I wanted to know what hand was guiding me along the Appalachian Trail. The answer, like the trail itself, expanded to planetary proportions and stretched back to earliest prehistory. I came to realize that I was being guided not by an invisible hand, but by a visible line: a path inscribed by trillions of living things, all setting forth, leading, following, veering off, connecting, finding shortcuts, and leaving their marks. The history of life on this planet can be seen as a single path made in the walking of it. We are all the inheritors of that line, but also its pioneers. Every step, we push forward into the unknown, following the path, and leaving a trail.

An old legend, passed down through the centuries, tells that Hanshan was last seen stepping into a crack in the face of Cold Mountain, which miraculously sealed shut behind him. At last, he had become one with the mountain he called home. All that remained of him were the lines he left behind.

AUTHOR'S NOTE

A s Whitman once wrote, this book is intended as "an exploration, as of new ground, wherein, like other primitive surveyors, I must do the best I can, leaving it to those who come after me to do much better." Everything in this book is, to the best of my knowledge, strictly factual. The order of the events, however, has been rearranged for clarity and narrative effect; the chapters, for example, do not flow in chronological order, nor do the sections. With any errors that you happen to encounter, please feel free to report them to robertmoor.ontrails@gmail.com, so that over time the book can continue to improve.

ACKNOWLEDGMENTS

E VERY BOOK, like every trail, is the result of a collaborative effort, but I would venture that this axiom is even truer of this book than most. Though I owe a debt of gratitude to hundreds of people who contributed to its completion, unfortunately, I only have space to thank a few. The elephant's share of thanks must go to Jon Cox, a truly tireless and perspicacious editor. Moreover, I am enormously grateful to have found my way to Bonnie Nadell, my agent, who shepherded this odd duck of a book from a fuzzy ball of ideas to a fully feathered, flight-ready manuscript. Also, to my publisher, Jon Karp, whose faith in my ability to complete this project seems, in hindsight, truly astonishing. To the other editors who have shaped pieces of this book, both large and small: Karyn Marcus, Robyn Harvie, Melissa Smith, Will Bleakley, and David Haglund. To my friend and sage legal counsel, Conrad Rippy. To my many advisors: Ted Conover, Robert Boynton, David Haskell, Robert Levine, Chris Shaw, Bill McKibben, Janisse Ray, and Rebecca Solnit. And to my friends, for their careful reading of an endless series of drafts over the years: Andrew Marantz, Sandra Allen, Will Hunt, and Ferris Jabr.

Throughout this process, I have reached out to hundreds of people with questions large and small. This is but a small sampling of those who have generously taken the time to respond. My thanks to: Iain Couzin, Jeff Lichtman, Ronald Canter, Steve Elkinton, Robin Sloan, Amy Lavender Harris, Matthew Tiessen, Elizabeth Barlow Rogers, Diana James, Sarah Wurz, Lawrence Buell, Robert Proudman, Jeffrey S. Cramer,

Signe Jeppesen, Reidun Lundgren, Toni Huber, Peter Cochran, Eustace Conway, E. O. Wilson, Judith Dupré, Stephen Budiansky, Ken Smith, Andy Downs, Jim Gehling, Ben Prater, Tracy Davids, Margaret Conkey, Basia Korel, Gary Snyder, Paolo Mietto, Bénédicte Jouneaux, Vincent Fourcassié, Lorraine Daston, Eva Williams, Mary Terrall, Rebecca Stott, Robert MacNaughton, Thomas Trott, Dan Rittschof, Marc Ratcliff, Charlotte Sleigh, Walter R. Tschinkel, John Bradley, Richard Bon, Ken Cobb, Eric Sanderson, Peter Coppolillo, Laurie Potteiger, Jordan Sand, Brandon Keim, Anthony Sinclair, Valerius Geist, Juliet Clutton-Brock, Jack Hogg, Fikret Berkes, Ted Belue, Andrew George, Martin Foys, J. Donald Hughes, Jason Neelis, Jennifer Pharr Davis, Justine Shaw, Jennifer Mathews, Rodney Snedeker, Carrie Gregory, James Cleland, Claudio Aporta, Thom Henley, Erick Leka, Lee Alan Dugatkin, Eric Johnson, A. J. King, Peter Devreotes, Rhonda Garelick, Bob Sickley, Peter Jensen, Arne Helgeland, Sivert Øgaard, Charlie Rhodarmer, Alexander Felson, and Bram Gunther.

Special thanks to: Virginia Dawson, Brett Leavy, Kelly Costanzo, Jake Stockwell, Simon Garnier, and Chris Reid.

Thanks to A. Laly, for her excellent work translating Bonnet's baroque French.

Thanks to the Middlebury Fellowship in Environmental Journalism, for helping fund the research for this book.

Thanks to Eva and Bob Morawski, Julia Morawski and Stephane Kowalczuk, Sue and Bill Guiney, Tami and Jerry McGee, Aysegul Savas and Maks Ovsjanikov, Louisa Bukiet and Peretz Partensky, David and Triss Critchfield, Ben and Emily Swan, and everyone else who housed and fed me during the many itinerant phases when this book was composed. Also, to all the drivers brave enough to pick a up mysterious stranger on the roadside and give him a ride to the next trailhead.

Finally, a great wave of love and gratitude to my family: Beverly, Bob, Alexis, Lindsay, Adrienne, Matt, Brook, all the Buntings, and all the rest. Thanks to Andy and Chris, who greeted me at the trail's end. And most of all, thanks to Remi, my co-adventurer, who always says yes to jumping in, no matter how cold the water.

ABOUT THE AUTHOR

Robert Moor has written for *Harper's, n+1, New York,* and *GQ,* among other publications. A recipient of the Middlebury Fellowship in Environmental Journalism, he has won multiple awards for his nonfiction writing. He lives in Halfmoon Bay, British Columbia.